Frontiers in Mathematics

Advisory Editorial Board

Benoît Perthame

Transport Equations in Biology

Birkhäuser Verlag
Basel · Boston · Berlin

Author:

Benoît Perthame
Département de Mathématiques et Applications
École Normale Supérieure, CNRS UMR 8553
45 rue d'Ulm
75230 Paris Cedex 05
France
e-mail: Benoit.Perthame@ens.fr

2000 Mathematical Subject Classification 35B30, 35F15, 35K55, 92D25, 92C17, 92C50

Library of Congress Control Number: 2006934214

Bibliographic information published by Die Deutsche Bibliothek
Die Deutsche Bibliothek lists this publication in the Deutsche Nationalbibliografie;
detailed bibliographic data is available in the Internet at <http://dnb.ddb.de>.

ISBN-13: 978-3-7643-7841-7 Birkhäuser Verlag, Basel – Boston – Berlin

© 2007 Birkhäuser Verlag, P.O. Box 133, CH-4010 Basel, Switzerland
Part of Springer Science+Business Media
Cover design: Birgit Blohmann, Zürich, Switzerland
Printed on acid-free paper produced from chlorine-free pulp. TCF ∞
Printed in Germany
ISBN-10: 3-7643-7841-7   e-ISBN-10: 3-7643-7842-5
ISBN-13: 978-3-7643-7841-7   e-ISBN-13: 978-3-7643-7842-4

9 8 7 6 5 4 3 2 1   www.birkhauser.ch

Contents

Preface

These lecture notes are based on several courses and lectures given at different places (University Pierre et Marie Curie, University of Bordeaux, CNRS research groups GRIP and CHANT, University of Roma I) for an audience of mathematicians. The main motivation is indeed the mathematical study of Partial Differential Equations that arise from biological studies. Among them, parabolic equations are the most popular and also the most numerous (one of the reasons is that the small size, at the cell level, is favorable to large viscosities). Many papers and books treat this subject, from modeling or analysis points of view. This oriented the choice of subjects for these notes towards less classical models based on integral equations (where PDEs arise in the asymptotic analysis), transport PDEs (therefore of hyperbolic type), kinetic equations and their parabolic limits.

The first goal of these notes is to mention (and describe very roughly) various fields of biology where PDEs are used; the book therefore contains many examples without mathematical analysis. In some other cases complete mathematical proofs are detailed, but the choice has been a compromise between technicality and ease of interpretation of the mathematical result. It is usual in the field to see mathematics as a black box where to enter specific models, often at the expense of simplifications. Here, the idea is different; the mathematical proof should be close to the 'natural' structure of the model and reflect somehow its meaning in terms of applications.

Dealing with first order PDEs, one could think that these notes are relying on the burden of using the method of characteristics and of defining weak solutions. We rather consider that, after the numerous advances during the 1980s, it is now clear that 'solutions in the sense of distributions' (because they are unique in a class exceeding the framework of the Cauchy-Lipschitz theory) is the correct concept. They allow for abstract manipulations, which we justify in the first section of the chapter 'General mathematical tools', and we use them freely throughout the text. Then one can concentrate on the intimate mathematical structure of the models.

It is a great pleasure for me to thank all those from whom these notes have profited; O. Diekmann who gave a series of enlightening lectures; my colleagues and collaborators and in particular J. Clairambault, L. Corrias and H. Zaag, who provided a constant motivation for better understanding; St. Boatto who made several useful suggestions; M. Desnous who helped me with figures. But mostly I would like to thank our postdocs at E.N.S. and former PhD students; most of the ideas emerged through discussions with them.

Paris, June 2006   Benoît Perthame

Chapter 1

From differential equations to structured population dynamics

Many problems arising in biology may be described, in a first formulation, using differential equations. This means that the model has been constructed by averaging some population and keeping only the time variable. This average usually hides some characteristic that can vary from one individual to another. Taking into account this characteristic leads to the so-called *structured population dynamic* equations. The aim of this chapter is to give several examples of this hidden characteristic and to show that it can sometimes represent a physical or physiological variable (ideally directly measurable). But sometimes it represents a biological or physiological variable that is not directly accessible to measurement but helps in a conceptual understanding of phenomena.

We give examples from *ecology*, which aims at understanding the relations between organisms themselves and their environment. We also give examples from *immunology*, which aims at understanding the interactions between hosts and parasites (a reference book on the mathematical aspects is [186]), and from the related field of *epidemiology*. These topics belong to the more general subject of *population biology*, see for instance [95, 218].

Many elaborate questions and mathematical tools (stability, bifurcation theory, Poincaré-Bendixson type theorems, index and topological concepts) can be used to study differential equations arising in biology. Examples can be found for instance in [217, 135, 133, 109]. Our purpose is not to present these mathematical aspects but to give examples where differential systems directly lead to Partial Differential Equations or Integral Equations. Therefore we only present some analysis when a simple argument can give a qualitative property of the system.

1.1   Invasions and space structure

The simplest model in population biology is the *unrestricted growth* of a population of size $N(t)$,

$$\frac{dN(t)}{dt} = \alpha N(t),$$

on which Th. Malthus ([165]) based his theory at the end of the 18th century. It represents an early stage of colonization in a virgin background where α represents, as we will see later, an available resource that determines the population growth. More elaborated is the *logistic growth* proposed by P.-F. Verhulst in the middle of the 19th century, where saturation occurs due to a maximum possible occupation density denoted by K,

$$\frac{dN(t)}{dt} = \alpha N(t)\Big(K - N(t)\Big).$$

Because $\alpha > 0$ here, the state $N = 0$ is unstable and the population density converges monotonically to $N(t = +\infty) = K$. In biology, the constant K is usually called the carrying capacity .

A further improvement in the 1950s (called Allee effect, [4]) is to suppose that too low densities $N(t)$ (less than K_- with the notation below) lead to extinction by lack of encounters between individuals. This can be modeled by the so-called bistable equation (the steady states $N(t) \equiv 0$ and $N(t) \equiv K_+$ are stable and $N(t) \equiv K_-$ is unstable)

$$\frac{dN(t)}{dt} = \alpha N(t)\Big(1 - \frac{N(t)}{K_-}\Big)\Big(\frac{N(t)}{K_+} - 1\Big), \qquad 0 < K_- < K_+, \quad \alpha > 0.$$

In fact, ODEs in mathematical modeling in biology go back to early infinitesimal calculus. For instance in 1760 Daniel Bernoulli [26] computed how many lives would be saved in Paris by inoculation against smallpop, establishing one of the first models of immunology.

A first natural question is how such equations can be established based on individual behaviors and why statistical effects can be neglected. This aspect, in the context of ecology, can be found in [183, 90, 91, 188] and in [101, 56] in the context of evolution theory.

A second natural question is what happens if the individuals can move in space. This leads to structuring the population by a space variable $x \in \mathbb{R}^d$ ($d = 2$ in practice, $d = 1$ is easier) and considering the population density $n(t, x)$ which at time t occupies the location x. Assuming a random motion of individuals (more precisely a diffusion process), we arrive at very famous equations. The logistic equation gives rise to the equation

$$\frac{\partial}{\partial t} n(t, x) - \Delta n(t, x) = \alpha n(t, x)\Big(K - n(t, x)\Big).$$

It is called the Fisher/KPP equation because it was introduced by Fisher ([105]) for the spread of a genetic trait, and studied by Kolmogorov, Petrovski and Piskunov ([153]). Biology is the prime motivation even though it is now very famous for studies of combustion and flame propagation).

The bistable equation is extended to become the so-called Allen–Cahn [1] equation (here $0 < \alpha < 1$ is a given parameter)

$$\frac{\partial}{\partial t} n(t,x) - \Delta n(t,x) = n(t,x)\big(1 - n(t,x)\big)\big(n(t,x) - \alpha\big). \tag{1.1}$$

It admits a traveling wave solution (front) with a speed that vanishes for $\alpha = .5$. Abusively — see (1.20) which propagates a pulse (or a spike) not a front — this equation is sometimes also called Fitzhugh–Nagumo equation [106, 182].

The study of these equations, and why they represent the propagation of an invasion front, has leds to numerous publications with probabilistic aspects ([110, 91]), study of traveling waves ([178, 25, 102, 188] and the references therein), or asymptotic methods ([17, 213, 199]). Also, Turing patterns ([220, 178]) are often obtained by a combination of such equations.

1.2   Ecology and Lotka–Volterra type of systems

1.2.1   The 2×2 Lotka–Volterra system

Mathematical ecology began with the model which is now known as the Lotka–Volterra system (see [229] for the original paper) or prey-predator system. Here $F(t)$ represents the population size of prey i.e. food for the predators $P(t)$. Prey are fed by the natural environment and thus grow with a rate α and are eaten by the predators at a rate βP (proportional to the number of predators). Predators die with a rate μ but their growth rate is proportional to the population size $F(t)$ which represents food for them. We arrive at the equations

$$\begin{cases} \dfrac{dF}{dt} = \alpha F - \beta F P, \\[2mm] \dfrac{dP}{dt} = \gamma P F - \mu P. \end{cases} \tag{1.2}$$

This model proposed by Volterra was successful for describing a specific observation on fishes in the Adriatic sea. The biologist D'Ancona observed the market for fish during the First World War when fishing had greatly decreased. According to (1.2), the market can be modeled by the steady state

$$\bar{F} = \frac{\mu}{\gamma}, \qquad \bar{P} = \frac{\alpha}{\beta}.$$

After the war, fishing increased and this led to the observation that the parameter α decreased by a small ratio, denoted by ε, while μ increased. The ratio of predatory fishes then changed according to

$$\left(\frac{\bar{P}}{\bar{F}}\right)_{\text{no fishing}} = \frac{\alpha}{\mu}\frac{\gamma}{\beta}, \qquad \left(\frac{\bar{P}}{\bar{F}}\right)_{\text{fishing}} = \frac{\alpha(1-\varepsilon)}{\mu(1+\varepsilon)}\frac{\gamma}{\beta} \approx \frac{\alpha}{\mu}\frac{\gamma}{\beta}(1-2\varepsilon). \qquad (1.3)$$

Therefore the model predicts that the proportion of predatory fishes is higher without fishing as observed during the war, in accordance with the observation of D'Ancona.

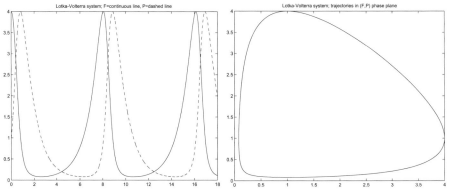

Figure 1.1: Solutions of Lotka–Volterra system with all parameters equal to 1; left: $F(t)$ (continuous line) and $P(t)$ (dashed line) for $0 \leq t \leq 18$, right: trajectories in (F, P) plane.

Another feature of this system is the possibility to understand easily the trajectories and we prove

Lemma 1.1. *For initial data $F(t=0) > 0$, $P(t=0) > 0$, the trajectories of system (1.2) are periodic and remain positive.*

Of course this qualitative lemma is in opposition to the above interpretation because it proves that the steady state is not locally attractive. This motivates us to introduce more elaborated (not merely quadratic) nonlinearities (see Section 1.2.2). Notice however that the quantitative conclusion is correct as one can see from the

Exercise. Prove that the average of $\left(F(t), P(t)\right)$ over one period is $(\bar{F}, \bar{P})$.
Hint: use the same change of unknown as in the proof of Lemma 1.1 below.

Proof of Lemma 1.1. Even though it is not hamiltionian, the Lotka–Volterra system can be reduced to a hamiltonian system by a change of unknowns, and thus it admits an invariant quantity. Consider the new variables (φ, ψ) and the hamiltonian $\mathcal{H}$,

$$\varphi = \ln F, \qquad \psi = \ln P,$$

$$\mathcal{H}(\varphi, \psi) = -\alpha\psi + \beta e^{\psi} + \gamma e^{\varphi} - \mu\varphi.$$

The system (1.2) becomes

$$\begin{cases} \dfrac{d\varphi}{dt} = \alpha - \beta P = \alpha - \beta e^{\psi} = -\mathcal{H}_{\psi}\big(\varphi(t), \psi(t)\big), \\[2mm] \dfrac{d\psi}{dt} = \gamma F - \mu = \gamma e^{\varphi} - \mu = \mathcal{H}_{\varphi}\big(\varphi(t), \psi(t)\big), \end{cases} \qquad (1.4)$$

(this is exactly the definition of a hamiltonian system). This hamiltonian structure implies that

$$\frac{d\mathcal{H}}{dt}(\varphi(t), \psi(t)) = \mathcal{H}_{\varphi}\frac{d\varphi}{dt}(t) + \mathcal{H}_{\psi}\frac{d\psi}{dt}(t) = 0.$$

In other words, for all times t we have $\mathcal{H}(\varphi(t), \psi(t)) = \mathcal{H}(\varphi(0), \psi(0))$ and thanks to the coercivity property

$$\mathcal{H}(\varphi, \psi) \to +\infty \quad \text{as} \quad |\varphi| + |\psi| \to \infty,$$

we deduce that the trajectories are indeed bounded in $(\varphi(t), \psi(t))$. This implies that $\big(F(t), P(t)\big)$ are bounded and bounded away from 0.

The periodicity follows from the same hamiltonian structure because trajectories are confined on the closed curves $\mathcal{H}(\varphi(t), \psi(t)) = \mathcal{H}(\varphi(0), \psi(0))$ and one readily checks that the velocity cannot vanish. A typical trajectory is depicted in Figure 1.1. □

In fact the terminology *Lotka–Volterra systems* nowadays contains the general class of differential systems of the form

$$\frac{dN_i(t)}{dt} = N_i(t)F_i\big(N_1(t), N_2(t), \ldots, N_I(t)\big), \qquad i = 1, 2, \ldots, I, \qquad (1.5)$$

where N_i denotes the density number of individuals of the species i, and F_i denotes the growth rate per capita of the species i. Among them is a generalization of the system (1.2), where

$$F_i(n_1, n_2, \ldots, n_I) = r_i + \sum_{j=1}^{I} a_{ij}n_j.$$

Another standard case is the *replicator system* which is again defined thanks to a square matrix A,

$$\frac{dN_i(t)}{dt} = N_i(t)\big((A.N)_i - N.A.N\big), \qquad A = A^t,$$

and has the property that solutions live on the subset $N_i \geq 0$, $\sum N_i = 1$. We refer to [136] for a survey on the state of the art for this type of these systems and relations to *game theory*, see also [135].

Most of the differential equations presented in this section are *Lotka–Volterra* systems.

1.2.2   Rosensweig–MacArthur

The periodicity property of solutions to the Lotka–Volterra system, related to its very particular form, makes it too specific because of the simplicity of the modeling. More realistic models have been derived which involve natural saturation effects. We present such a prey-predator model now, taken from [204], which describes the population density $F(t)$ of algae (food) and the density $P(t)$ of *Daphnia* (a waterflea, the predator).

$$\begin{cases} \dfrac{dF}{dt} & = \alpha F(1 - \frac{F}{K}) - I_{max}\frac{F}{F_h+F}P, \\[3mm] \dfrac{dP}{dt} & = \bar{\gamma}I_{max}\frac{F}{F_h+F}P - \mu P. \end{cases} \qquad (1.6)$$

Here, α and K denote the maximum growth rate and the carrying capacity of the algae, I_{max} and F_h are the maximum feeding rates and half-saturation food level of the *Daphnia* functional response; $\bar{\gamma}$ and μ are the conversion efficiency and per capita mortality rate of *Daphnia*. All these parameters are positive.

Lotka–Volterra is a limiting case of this system when we take the parameters as

$$K, I_{max}, F_h \to \infty, \qquad \frac{I_{max}}{F_h} \to \beta, \qquad \gamma = \beta\,\bar{\gamma}.$$

In this limit, the trajectories of the Rosensweig–MacArthur system converge to those of the Lotka–Volterra system thanks to the general theory of continuity with respect to parameters in Cauchy-Lipschitz theory (see for instance [65]).

Of course, such a system preserves the positive cone $F(t) > 0$ and $P(t) > 0$ and the upper bound $F(t) \leq K$. We now assume that the parameters satisfy

$$\mu < \bar{\gamma}I_{max}, \qquad \bar{F} < K,$$

where $(\bar{F}, \bar{P})$ is the unique positive steady state with $\bar{F}$ uniquely defined thanks to

$$\frac{\bar{F}}{F_h + \bar{F}} = \frac{\mu}{\bar{\gamma}I_{max}},$$

$$\bar{P} = \frac{F_h + \bar{F}}{I_{max}}\,(1 - \frac{\bar{F}}{K}).$$

It is easy to check that this steady state is attractive: all trajectories, with $F(0) > 0$, $P(0) > 0$ converge to $(\bar{F}, \bar{P})$ as $t \to +\infty$. A typical trajectory is presented in Figure 1.2.

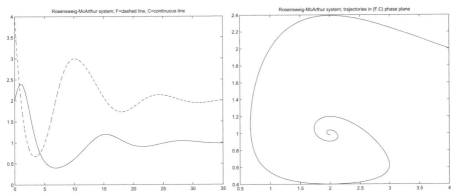

Figure 1.2: SOLUTIONS OF ROSENSWEIG–MACARTHUR SYSTEM WITH ALL PARAMETERS EQUAL TO 1 EXCEPT $K = 4$, $I_{max} = F_h = 2$; LEFT: $F(t)$ (DASHED LINE) AND $P(t)$ (CONTINUOUS LINE) FOR $0 \leq t \leq 35$, RIGHT: TRAJECTORIES IN (F, P) PLANE.

1.2.3 Chemostat (1); several nutrients

A particularly representative differential system of ecology arises in laboratory experiments. A *chemostat* (see also Section 1.6) contains nutrients S_i, $i = 1, 2, \ldots, I$, and a micro-organism (the example of *Daphnia* is frequent, see also Section 4.3.1) which uses the nutrients to grow. The modeling is particularly simple because $S_i(t)$ can measure the mass of chemical constituents that are either free in the chemostat or absorbed by the micro-organism whose biomass is denoted by $n(t)$. Therefore we can write an exact balance equation

$$\begin{cases} \dfrac{d}{dt} S_i(t) & = R[S_{0i} - S_i(t)] - S_i(t)\eta_i n(t), \\[2mm] \dfrac{d}{dt} n(t) & = n(t) \left(\displaystyle\sum_{i=1}^{I} S_i(t)\eta_i - R \right). \end{cases} \qquad (1.7)$$

Here the vector with positive coefficients $(S_{0i})_{i=1,\ldots,I}$ represents the influx of 'pure' nutrients and the terms $-RS_i(t)$ and $-Rn(t)$ represent the outflux of the mixture with rate $R > 0$. The quadratic term, as usual, represents the predation, with rate $\eta_i > 0$, of the constituent S_i.

As we mentioned before, this system contains a fundamental balance law for the total mass $M(t)$ of constituents (free or absorbed) defined by

$$M(t) = \sum_{i=1}^{I} S_i(t) + n(t),$$

$$\frac{d}{dt} M(t) = R(\sum_{i=1}^{I} S_{0i} - M(t)), \qquad \forall t \geq 0. \qquad (1.8)$$

Several more elaborated models exist and many questions are of interest for applications in ecology (limiting constituents, degradation of $n(t)$ after death...), cf. [70, 203, 204].

As a mathematical result, let us just mention that the steady state $(\bar{n}, \bar{S}_1, \ldots, \bar{S}_I)$ is easy to find. If $\bar{n}$ does not vanish, it is given by

$$RS_{0i} = \bar{S}_i[R + \eta_i\bar{n}], \qquad \sum_{i=1}^{I} \bar{S}_i\eta_i = R,$$

which can be solved uniquely (by monotonicity in $\bar{n}$) as

$$\bar{S}_i = \frac{RS_{0i}}{R + \eta_i\bar{n}}, \qquad \sum_{i=1}^{I} \bar{S}_i\eta_i = R. \tag{1.9}$$

Of course this is only possible if $\sum_{i=1}^{I} S_{0i}\eta_i > R.$

Lemma 1.2. *Consider an initial state $n(0) > 0$, $S_i(0) \geq 0$ and assume that $\sum_{i=1}^{I} S_{0i}\eta_i < R$. Then the solution to (1.7) satisfies*

$$n(t) \xrightarrow[t\to\infty]{} 0, \qquad S_i(t) \xrightarrow[t\to\infty]{} \bar{S}_{0i}. \tag{1.10}$$

Lemma 1.3. *Consider an initial state $n(0) > 0$, $S_i(0) \geq 0$ and assume that $\sum_{i=1}^{I} S_{0i}\eta_i > R$. Then the solution to (1.7) satisfies*

$$n(t) \xrightarrow[t\to\infty]{} \bar{n}, \qquad S_i(t) \xrightarrow[t\to\infty]{} \bar{S}_i, \tag{1.11}$$

and $\bar{n}$, $\bar{S}_i$ are uniquely given (implicitly) by (1.9).

Notice that the model expresses a balance law. From (1.8), we deduce that the quantity $M(t) = \sum_{i=1}^{I} S_i(t) + n(t)$ satisfies

$$M(t) = \Sigma - Q^0 e^{-Rt}, \qquad Q^0 = M(0) - \Sigma, \qquad \Sigma = \sum_{i=1}^{I} S_{0,i}. \tag{1.12}$$

Such a bound implies that there are unique global solutions using the Cauchy–Lipschitz theorem.

Proof of Lemma 1.2. The equation for S_i can be written

$$\frac{d}{dt}[S_i(t) - S_{0i}] \le -R[S_i(t) - S_{0i}],$$

and thus we have also found a Lyapunov functional,

$$\frac{d}{dt}[S_i(t) - S_{0i}]_+ \le -R[S_i(t) - S_{0i}]_+,$$

which proves that

$$\text{limsup}_{t\to\infty}\ S_i(t) \le S_{0i}, \qquad \text{limsup}_{t\to\infty}\sum_{i=1}^{I}\eta_i S_i(t) \le \sum_{i=1}^{I}\eta_i S_{0i} < R.$$

Therefore the growth rate of n, namely $\sum_{i=1}^{I}\eta_i S_i(t) - R$ is negative for times large enough and thus $n(t)$ vanishes exponentially. Then, we deduce from the *limsup* of S_i and (1.12) that in fact $S_i(t) \to S_{0i}$. $\square$

Proof of Lemma 1.3. We only prove that $\text{liminf}_{t\to\infty}n(t) \ge \bar{n}$ and leave the reverse inequality to the reader.

Firstly, we notice that from the proof above, we have

$$\text{limsup}_{t\to\infty}S_i(t) \le S_{0i}. \tag{1.13}$$

Then, we begin by proving that

$$\text{liminf}_{t\to\infty}n(t) \ge n_1 := \frac{\sum \eta_i S_{0i} - R}{\sum \eta_i} > 0. \tag{1.14}$$

Otherwise, we have $\forall \varepsilon, \exists t(\varepsilon)$ such that for $t > t(\varepsilon)$ we have $n(t) < n_1 - 2\varepsilon$. And from (1.12), we deduce that for t large enough

$$\sum S_i(t) \ge \Sigma - n_1 + \varepsilon,$$

and then it follows that

$$S_i(t) \ge S_{0i} - n_1 + \varepsilon, \qquad \forall i = 1, \dots, I,$$

(otherwise i with $S_i(t) \le S_{0i} - n_1 + \varepsilon$ and (1.13) contradicts the previous statement).

As conclusion, we have

$$\sum \eta_i S_i(t) \ge \sum \eta_i S_{0i} - n_1 \sum \eta_i + \varepsilon \sum \eta_i = R + \varepsilon \sum \eta_i.$$

This implies an exponential growth of $n(t)$ which contradicts the boundedness of $M(t)$ and proves (1.14).

From this we also deduce that

$$\mathrm{limsup}_{t\to\infty} S_i(t) \leq \frac{RS_{0i}}{R + \eta_i n_1}.$$

Indeed, if for some i and t_1 we have $S_i(t_1) > \frac{RS_{0i}}{R+\eta_i n_1}$, then from the equation of S_i in (1.7), we deduce that $S_i(t_1)$ decreases.

Then, we can go further and deduce (in the case $n_1 < \bar{n}$ we are done),

$$\mathrm{liminf}_{t\to\infty} n(t) \geq n_2 := n_1 \sum \frac{\eta_i S_{0i}}{R + \eta_i n_1} > n_1.$$

This inequality is again a consequence of (1.13). It remains to iterate the argument and prove that

$$\mathrm{limsup}_{t\to\infty} S_i(t) \leq \frac{RS_{0i}}{R + \eta_i n_k}, \quad \mathrm{liminf}_{t\to\infty} n(t) \geq n_{k+1} := n_k \sum \frac{\eta_i S_{0i}}{R + \eta_i n_k} > n_k.$$

As $k \to \infty$ we arrive at the conclusion.   □

Exercise. Consider the following variant of the several nutrients chemostat equation (with conversion factors γ_i) and prove a result similar to Lemma 1.3,

$$\begin{cases} \dfrac{d}{dt}S_i(t) &= S_{0i} - S_i(t) - S_i(t)\eta_i n(t), \\[2mm] \dfrac{d}{dt}n(t) &= -n(t) + \displaystyle\sum_{i=1}^{I} \gamma_i S_i(t)\eta_i n(t). \end{cases} \tag{1.15}$$

Hint: Change $S_i(t)$ in $\mu_i S_i(t)$ and choose μ_i properly.

1.3 Hodgkin–Huxley, Fitzhugh–Nagumo and pulse propagation

Electric models with ODE were used very early in physiology. The first one is maybe the van der Pol system ([222, 223]) for heartbeat. The Hodgkin–Huxley system has remained as the first model for the propagation of nerve impulses, along the giant squid axon in the original experiments during the 1950s (see [178, 150] and the references therein). These models initiated the theories of electrophysiology and cardiac or neural rhythms (see [120]). Signal regulation, bursting waves (episodic bursts of high frequency oscillations), pulses (spikes) propagation enter the biological observations made in these areas that sustain the modeling proposed by Hodgkin–Huxley and simplified by Fitzhugh and Nagumo [106, 182]. We refer to [11, 89, 133, 109] for the explanation of auto-oscillations and bursting oscillations. Here we just explain the slow-fast dynamics exhibited in this system and the dynamics of pulses.

1.3.1 Van der Pol equation and auto-oscillations

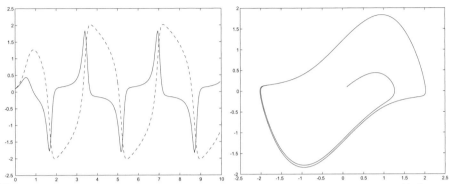

Figure 1.3: Solutions of the van der Pol system (1.17) with $A = 5$; left: $x(t)$ (continuous line) and $y(t)$ (dashed line), right: trajectories in the phase plane (x, y).

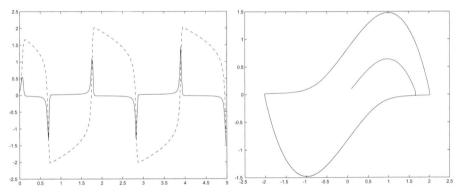

Figure 1.4: Solutions of the van der Pol system (1.17) with $A = 40$; left: $x(t)$ (continuous line) and $y(t)$ (dashed line), right: trajectories in the phase plane (x, y).

The van der Pol equation is

$$
\begin{cases}
\dot{x}(t) & = Ay, \\
\dot{y}(t) & = -x + Ay(1 - x^2).
\end{cases}
\tag{1.16}
$$

Here we have used an unusual scaling which is compatible with the limit $A \to \infty$ which is the case of real interest.

Usually the case when A is small is treated. Then, one can better see the behavior on rescaled unknowns $X(t) = x(t/\sqrt{A})$, $Y(t) = \sqrt{A}\, y(t/\sqrt{A})$ and it reduces to $\dot{X}(s) = Y(s)$, $\dot{Y}(s) = -X(s) + o(A)$, i.e., the harmonic oscillator whose solutions, $(X, Y) = z_0\, e^{is}$, are periodic with energy $E(s) = E^0 = (X^0)^2 + (Y^0)^2$. This expresses that a slow dynamics corresponding to the term $o(A)$ corrects the fast dynamics $\dot{X}(s) = Y(s)$, $\dot{Y}(s) = -X(s)$. The reader will easily find additional matter on this subject in [11, 133, 109].

When $A > 0$ the situation changes drastically and trajectories are attracted by a single stable limit cycle (periodic trajectory). A qualitatively different phenomena occurs called *auto-oscillations* or *self-excited oscillations*. This phenomena is exhibited in Figures 1.3 for $A = 5$ and 1.4 for $A = 40$. As one can see, the natural limit here is indeed $A \to \infty$ which we explain in the next subsection and which provides another, and more interesting, example of slow-fast dynamics.

1.3.2 Slow-fast dynamics and electric pulses

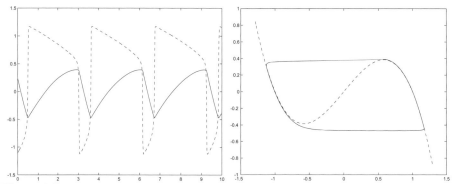

Figure 1.5: SOLUTIONS OF THE FITZHUGH–NAGUMO EQUATION (1.17) WITH $x_1 = .55$, $\varepsilon = 0.01$; LEFT: $x(t)$ (DASHED LINE) AND $y(t)$ (CONTINUOUS LINE), RIGHT: TRAJECTORIES IN THE PHASE PLANE (x, y) (CONTINUOUS LINE) AND THE CURVE $y = x(1 - x^2)$ (DASHED LINE).

As we have seen, the van der Pol system is an example of *slow-fast dynamics* and the same concept applies to the Fitzhugh–Nagumo equation

$$\begin{cases} \dot{x}(t) &= \frac{1}{\varepsilon}[x(1 - x^2) - y], \\[2mm] \dot{y}(t) &= x - x_1, \end{cases} \tag{1.17}$$

where $\varepsilon > 0$ is a (small) parameter and $x_1 \in \mathbb{R}$ is a crucial parameter to be chosen later. This system, that extends the bistable equation (see Section 1.1), describes an electric potential and $x(t)$ does not stay within the interval $[-1, 1]$. Notice that it can be derived from the van der Pol system (1.16) by Lienard's change of variable.

The cubic null-cline $y = f(x) = x(1 - x^2)$ achieves a local maximum at $x_0 = 1/\sqrt{3}$ and we set $y_0 = f(x_0)$. Since f is odd, $-x_0$ is a local minimum point and $-y_0 = f(-x_0)$. Therefore, the function f can be inverted in the two stable branches as follows:

$$
\begin{cases}
x \in]-\infty, -x_0], & y = f(x) \iff x = G_+(y), \quad y \in]-\infty, -y_0], \\
x \in [x_0, \infty[, & y = f(x) \iff x = G_-(y), \quad y \in [y_0, \infty[.
\end{cases}
\tag{1.18}
$$

When $x_1 \in]-x_0, x_0[$, the system exhibits periodic solutions (again auto-oscillations) as depicted in Figure 1.5.

For ε small, we can see that the solution approaches one of the two branches $x = G_\pm(y)$ and the system is reduced to $\dot{y} = G_\pm(y) - x_1$ along these two branches, with a jump from G_+ to G_- when $y(t)$ exceeds the value $f(x_0)$ (and vice-versa). Again, we refer to [11, 133, 109] for additional matter.

1.3.3 Fitzhugh–Nagumo and concentration pulses

Another type of Fitzhugh–Nagumo system serves to describe concentration pulses. Therefore we now impose that $x(t) \geq -1$ in the following variant of (1.17) where $x(t) + 1$ is a concentration:

$$
\begin{cases}
\dot{x}(t) = \frac{1}{\varepsilon}[x(1 - x^2) - (1 + x)y], \\
\dot{y}(t) = [\psi_\infty(x) - y].
\end{cases}
\tag{1.19}
$$

The function ψ_∞ is chosen, for instance, as

$$
\psi_\infty(x) = \begin{cases}
0 & \text{for } x \leq 0, \\
\text{increasing} & \text{for } 0 \leq x \leq 1.
\end{cases}
$$

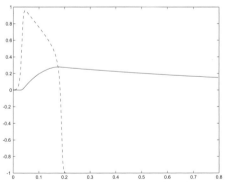

Figure 1.6: Solutions of Fitzhugh–Nagumo system (1.19) with $\varepsilon = .005$, $x^0 = 10^{-3}$; $x(t)$ (dashed line) and $y(t)$ (continuous line).

It is used to describe a (say calcium) wave. Here $-1 \leq x(t) \leq 1$, $y(t) \geq 0$ is an invariant region. The state $x = -1$ refers to normal state and $x = 1$ refers to depressed cells. These are the two stable states for the equation on x when y is taken to be 0. The relaxation function $\psi_\infty(\cdot)$ plays the role of relaxing depressed cells to their normal state with constant of time large compared to the dynamics of $x(t)$. With an appropriate choice of ψ_∞, there is a single stable state $\bar{x} = -1$, $\bar{y} = 0$. Typically an outside instability deviates x^0 from the stable state to an initial value $x^0 > 0$. This drives the system to the stable state $x \approx 1$ until $y(t)$, with some delay due to τ which is chosen large enough, compensates for it and brings back the system to its stable state. In Figure 1.6, the function $\psi_\infty(x) = 4\,x^2$ is used for $x \geq 0$.

1.3.4   Fitzhugh–Nagumo system with space structure

Once again, it is natural to consider a space structured version of the above type of Fitzhugh–Nagumo system. It is used to describe wave (spikes, pulses) propagations in a spatial region, as potential waves in neural communication by electrical signaling or calcium waves in the brain during vascular cerebral incidents (see [123] for instance).

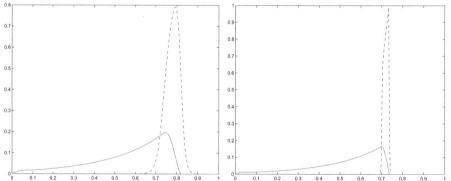

Figure 1.7: Solutions of the space structured Fitzhugh–Nagumo system (1.20) with $\alpha = .2$ and $\varepsilon = 0.01$ (left), $\varepsilon = 0.001$ (right); we represent both $u(t, x)$ (dashed line) and $v(t, x)$ (continuous line) as a function of x, at a given time. The pulse propagates from left to right.

We change notation now and denote by $n(t, x)$, $v(t, x)$ (they were denoted by $x + 1$ and y in the previous sections) the solutions to

$$
\begin{cases}
\frac{\partial}{\partial t} n(t, x) - \varepsilon \Delta n & = \frac{1}{\varepsilon} n[(1 - n)(n - \alpha) - v], \\[2mm]
\frac{\partial}{\partial t} v(t, x) & = v_\infty(n) - v.
\end{cases}
\tag{1.20}
$$

Notice also the analogy with the Allen–Cahn equation for invasion fronts (1.1) in Section 1.1. Figure 1.7 shows a traveling wave solution (pulse, spike) with $v_\infty(n) = 3.(n - .4)_+$ (here the numbers are adapted to $\alpha = .2$). We choose here

$\alpha = .2$ in order to propagate a wave on n in the Allen–Cahn equation and the equation on v, with a certain delay inhibiting the pulse back to the other stable value $n = 0$,

1.4 Virus dynamics and the immune system

A wide class of Lotka–Volterra type of systems arises in the study of host-parasite dynamics. A typical example is virus/immune system competition but epidemiology is closely related too. We now explain how differential systems can be extended to structured population systems in the case of virus dynamics. The next section treats epidemiology. The intent is mainly to give examples where the structuring variable has a physiological meaning (these are referred to as *physiologically structured*), besides the traditional space structure already mentioned for invasion fronts in Sections 1.1 and 1.3.

1.4.1 Virus dynamics, immune response

Describing the immune response of a single individual is a domain full of differential equations. A simple idea is that virusses represent the prey and leukocytes are the predators, then a model like the prey-predator system of Section 1.2.2 is satisfactory. However many other effects have to be added.

One of the most general and basic facts in this domain is that not only the free virus population size $v(t)$ and the population size of uninfected cells $f(t)$ (food for virus) has to be considered, but also infected cells $c(t)$. Indeed, new viruses are only produced by the infected cells (they enter the cell and change a segment of DNA which then reproduces the modified code). On the other hand a free virus can infect an uninfected cell. We arrive at a system

$$\begin{cases} \dfrac{df}{dt} &= \alpha - \mu_f f - \beta f v, \\[2ex] \dfrac{dc}{dt} &= \beta f v - \mu_c c, \\[2ex] \dfrac{dv}{dt} &= \gamma c - \mu_v v. \end{cases} \tag{1.21}$$

Consequences of this simple description, further models and comparisons to in vivo data can be found in [186], [82] and the references therein.

1.4.2 Virulence structure

A major feature has been neglected in the description of Section 1.4.1. Namely, the possible mutations of a virus which continuously changes its 'shape', i.e., its molecular conformation. A general account of these aspects can be found in the survey paper [195]. In [186], it is proposed to take this into account by a large (but finite) system such as (1.21); in [112] a continuous model with mutations is proposed with a model close to adaptive dynamics as studied in Section 2.4 below.

In the case of the competition between a tumor and the immune system, a population structured by a virulence parameter is proposed in [74, 73] (see also the references therein). The main effect is that the new tumor cells escape continuously from the attack of the immune system and increase their agressivity. It can also be thought as very fast and asymmetric mutations (compared to Section 2.4), see [112]. In a very simplified case, the model can be written as two equations on the densities $n_1(t, x)$ of tumor cells with activation $x \geq 0$ and $n_2(t, x)$ of immune system cells with defense ability $x \geq 0$,

$$
\begin{cases}
\frac{\partial}{\partial t} n_1(t, x) + \frac{\partial}{\partial x}[x(A_1(t) - A_2(t))n_1] & = [\alpha_1 x - \beta A_2(t)]n_1, \\[2mm]
\frac{\partial}{\partial t} n_2(t, x) + \frac{\partial}{\partial x}[-\gamma x A_1(t) n_2] & = \alpha_2 x A_1(t) n_2, \\[2mm]
A_i(t) & = \int_{\{x \geq 0\}} x\, n_i(t, x) dx.
\end{cases}
\tag{1.22}
$$

The modeling assumptions here are that there is a continuous source of nutrient for both tumor and immune system cells with rate α_i. The immune system kills tumor cells with a rate β, and also prevents their natural progression through the $A_2(t)$ term in the x-derivative arising in the first equation. Meanwhile the tumor cells depress the immune system through the $A_1(t)$ term in the x-derivative arising in the second equation.

The long time behavior of this system is studied in [73, 77] (with a third equation for the normal cells, a nutrient for the tumor cells). The outcome is that, depending on the initial data, either the tumor cells can 'win' ($A_1(t) \to \infty$), or the immune system wins and $A_1(t) \to 0$. But there are several ways to express more precisely the long time asymptotic and several points are still open. A specific difficulty comes from the multiplication by x in several terms; this induces growth at infinity which makes the various estimates more difficult.

1.5   Models of epidemiology

We give here another example of a physiologically structured population model in the context of epidemic propagation. Again the structure variable represents a biological feature.

1.5.1   Discrete states; SI, SIR, SEIR

The question in epidemiology is to describe the effect of virus infection on a population and no longer on a single individual as in Section 1.4.1. Simplifying again as much as possible, one is led to define the number density per surface unit of various populations. As a first model one can consider susceptible individuals with size $S(t)$, and infective population with size $I(t)$. When the birth rate, B below, is constant and limited to susceptible newborn (as the simpler possible model), we arrive at the SI model

$$\begin{cases} \dfrac{dS}{dt} = B - \mu_S S - \gamma_S S\, I, \\[2mm] \dfrac{dI}{dt} = \gamma_S S\, I - \mu_I I. \end{cases} \tag{1.23}$$

Notice the striking analogy with the Lotka–Volterra system (1.2) and the chemostat system (1.2.3). Here the susceptible individuals get infected by contact with other infected individuals with a rate γ_S, therefore giving the quadratic loss term $S\, I$ in the first equation and the corresponding gain term in the second equation. The susceptible individuals death rate is denoted by μ_S and the infected population death rate is denoted by $\mu_I > \mu_S$. The total population is therefore $N = S + I$.

In a next step one can also consider that a number of infected individuals recover, and then are immune. We denote this population density by $R(t)$ (for Removed). Then, in this class of models called (SIR), an example is

$$\begin{cases} \dfrac{dS}{dt} = \beta_S(S + I + R) - \mu_S S - \gamma_S S\, I, \\[2mm] \dfrac{dI}{dt} = \gamma_S S\, I - \mu_I I - \beta_R I, \\[2mm] \dfrac{dR}{dt} = \beta_R I - \mu_R R. \end{cases} \tag{1.24}$$

Here the interpretation is that uninfected individuals are born from the total population and die with a rate μ_S. They get infected by contact with other infected individuals with a rate γ_S, therefore giving the quadratic loss term $S\, I$ in the first equation and the corresponding gain term in the second equation. Finally, individuals of the infected population die with a rate μ_I (which in principle is higher than the other mortality rates μ_S and μ_R) and can recover with a rate β_R. This gives rise to the gain term $\beta_R I$ for the recovered population. Here the total population is $N = S + I + R$.

Of course, at this level, the model shares many similarities with the immunology system of Section 1.4.1. This comes from the common feature that they always describe host-parasite populations. The values of the parameter differ however drastically from one situation to the other and also the extensions to more realistic models (see [184, 218, 82, 186]).

The main question in this area is to know if an epidemic will spread or not. This is related to the *basic reproduction number*, classically denoted by R_0 and defined as the number of secondary cases from a single infected individual in a 'virgin population' with $\bar{I} = \bar{R} = 0$, $\bar{S} = \mu_S/\beta_S$. In the case of (1.24), one can easily check that $R_0 = S(0)\gamma_S/(\mu_I + \beta_R)$.

Other typical questions in this area are: to know wether the steady states of such a system are stable, to know the effect of vaccination on the global population, to add spatial effects and stochasticity, to take into account other intermediary states (continuous variables are then usually considered) and that immunity can be lost after some time. Also a classical effect leads to the so-called SEIR models (E for exposed, those individuals who catch the disease and are in the latent period, they are infected but not yet infectious).

1.5.2   The Kermack–McKendrick model

In this situation, it is of course natural to postulate models where the infection state is continuous. This has been proposed by several authors beginning with Kermack and McKendrick [162, 151]. The book [82] is a very good and recent account of the subject, see also the analysis of the system below in [143, 231].

$$\frac{d}{dt}S(t) = B - \mu_S S(t) - \lambda_S(t)S(t),$$

$$\lambda_S(t) = \int_0^\infty \kappa(x)n(t,x)dx,$$

$$\begin{cases} \dfrac{\partial}{\partial t}n(t,x) + \dfrac{\partial}{\partial x}n(t,x) + (\mu_I + \beta_n(x))n(t,x) = 0, \\[2mm] n(t,x=0) = \lambda_S(t)S(t), \end{cases}$$

$$\frac{d}{dt}R(t) = \int_0^\infty \beta_n(x)n(t,x)dx.$$

The interpretation is that $n(t,x)$ is the density of population at age x of infection and this replaces the compartment $I(t)$ in the SIR system (1.24). Individuals are infected (at age $x = 0$ in the infection stage) from susceptible individuals that are infected by encounter with the infected population at a rate $\lambda_S(t)$ depending of the time x elapsed since infection through a rate $\kappa(x)$. Notice again the quadratic term for transition from susceptible to infected. The advantage of the Kermack–McKendrick model is that one can take into account the variable infectivity level, and removal rate, depending on the age of the disease. A variant would be to consider that infected are Removed at a certain age $x_\sharp$.

1.5.3   An immunity structured model

We would like to also mention another physiologically structured model for the spread of myxoma virus among a rabbit population that we took from [232].

These authors take into account encounters with infected and uninfected rabbits in a more detailed manner because susceptibles have a certain immunity level, in opposition to the Kermack–McKendrick model. They arrive at the system

$$
\begin{cases}
\frac{\partial n(t,x)}{\partial t} - \frac{\partial}{\partial x}(v_n(x)n(t,x)) = (\beta_n - \mu_n)n(t,x) - \gamma \int_0^1 K(x,x')n(t,x)p(t,x')dx', \\[2mm]
\frac{\partial p(t,x)}{\partial t} - \frac{\partial}{\partial x}(v_p(x)p(t,x)) = -\mu_p p(t,x) + \gamma \int_0^1 K(x,x')n(t,x)p(t,x')dx'.
\end{cases}
$$
$$(1.25)$$

Here n represents the population of uninfected rabbits with an immunity level x. They are born with a rate β_n and become infected by encountering infected individuals with a rate γ. Also $p(t,x)$ represents the population of infected rabbits and they only come, by infection, from the n population. As usual μ_n and μ_p denote the mortality rates in the two populations.

It can be considered as a drawback to structure the population with a parameter that is not physically measurable (as the immunity level x here) and age is supposed to be an easier parameter [108].

1.6 Ecological model of competition for resources

1.6.1 Chemostat (2): several micro-organisms

We come back to the ecology problem of the chemostat already mentioned in Section 1.2.3. We now consider a culture of several micro-organisms in a single substrate (see Figure 1.8). In this situation, the system is given by

$$
\begin{cases}
\dot{S}(t) = R(S_0 - S) - \displaystyle\sum_{j=1}^{I} \eta_j(S)N_j, \\[4mm]
\dot{N}_i(t) = N_i\left(\eta_i(S) - R\right), \\[2mm]
S(0) = S^0 > 0, \quad N_i(0) = N_i^0 > 0, \ \forall i = 1,\ldots,I.
\end{cases}
$$
$$(1.26)$$

The variable $S(t)$ denotes the single substrate (in terms of the mass of a given representative constituent), $N_i(t)$ is the biomass of the i-th micro-organism in the chemostat (written in terms of the same constituent), R is the dilution rate of the input flow of nutrient concentration S_0. Finally, the ability for the i-th organism to use the nutrient S depends only on S in the simplest model, and is denoted by $\eta_i(S)$ (uptake rates).

Several variants of the system (1.26) have been proposed. In particular one can consider that intraspecific competition for the resource leads to uptake functions $\eta_j(S, N_j)$ that decrease with N_j. We refer to [159, 160] for the analysis of such cases and for further models.

Figure 1.8: THE PRINCIPLE OF THE CHEMOSTAT. AN INFLOW OF PURE NUTRIENT S_0, WITH RATE R, IS COMPENSATED BY AN OUTFLOW CONTAINING BOTH THE MICRO-ORGANISMS AND THE NUTRIENT AT CONCENTRATION $S(t)$.

One usually assumes that,

$$\eta_i(\cdot) \text{ is increasing}, \qquad \eta_i(S_0) > R, \quad \forall i = 1, \ldots, I, \tag{1.27}$$

$$\text{the numbers } \eta_i^{-1}(R) \quad \text{are all different.} \tag{1.28}$$

Then there are $I + 1$ steady states, the trivial one $N_i \equiv 0$, $S = S_0$, and those I states composed of a single micro-organism

$$(0, \ldots, 0, \bar{N}_i, 0, \ldots, 0), \qquad \bar{S} = \eta_i^{-1}(R) < S_0, \qquad \bar{N}_i = S_0 - \eta_i^{-1}(R).$$

One can prove the following result, a first example of the selection principle

Theorem 1.1. *Under assumptions* (1.27), (1.28), *among these steady states, the specific i_0 corresponding to $S^* = \min_{1 \le i \le I} \eta_i^{-1}(R)$ (largest population) gives the globally asymptotically stable state. In other words, one has $N_i(t) \to 0$ for $i \ne i_0$, $N_{i_0}(t) \to N_{i_0}^* = S_0 - S^*$, $S(t) \to S^*$ as $t \to \infty$.*

Remark 1.1. In the case when $\eta_i(\cdot)$ are linear, $\eta_i(S) = \eta_i^* S$, then $\eta_i^{-1}(R) = R/\eta_i^*$ and the index i_0 also corresponds to the formula $\eta_{i_0}^* = \max \eta_i^*$.

Proof. We follow ideas already used for the chemostat with several nutrients but the additional nonlinearity in the uptake rates leads to improvement of the method. We divide the proof in four steps. We first consider a balance law of the system, then we prove a lower bound on $\sum N_i(t)$, we then prove that S has a limit and then conclude that $S(t) \to S^*$ as $t \to \infty$.

First step. A balance law. Adding the two equations on S and N_i, we have

$$\frac{d}{dt}[S(t) + \sum_{i=1}^{I} N_i(t)] + R[S(t) + \sum_{i=1}^{I} N_i(t)] = RS_0.$$

Therefore

$$S(t) + \sum_{i=1}^{I} N_i(t) = S_0 + Q_0 \, e^{-Rt}, \qquad Q_0 = [-S_0 + S^0 + \sum_{i=1}^{I} N_i^0]. \tag{1.29}$$

We also notice that

$$S(t) > 0, \quad N_i(t) > 0, \qquad \forall t \geq 0, \ \forall i = 1, \ldots, I.$$

Second step. A lower bound (non-extinction). We prove that

$$\liminf_{t \to \infty} \sum_{i=1}^{I} N_i(t) = \underline{M} > 0.$$

Indeed, we have, from the equation on N_i,

$$\frac{d}{dt} \sum_{i=1}^{I} N_i(t) \geq \left(\underline{\eta}(S) - R \right) \sum_{i=1}^{I} N_i(t), \qquad \text{with} \quad \underline{\eta}(S) = \inf_{1 \leq i \leq I} \eta_i(S).$$

If $\sum_{i=1}^{I} N_i(t)$ approaches 0, $S(t)$ approaches S_0 (using the step 1), and thus $\sum_{i=1}^{I} N_i(t)$ increases, using assumption (1.27).

Third step. Asymptotic of S. We now prove that $S(t)$ has a limit as $t \to \infty$. To do that we compute the equation on $\dot{S}(t)$. Using that, from the first step,

$\dot{S} + \sum_{i=1}^{I} \dot{N}_i = -RQ_0 e^{-Rt}$, we have

$$
\begin{aligned}
\frac{d}{dt} \dot{S}(t) &= -\sum_{i=1}^{I} \eta_i(S)\dot{N}_i - \dot{S} \sum_{i=1}^{I} \eta_i'(S)N_i - R\dot{S} \\
&= -\sum_{i=1}^{I} \left(\eta_i(S) - R \right)\dot{N}_i - \dot{S} \sum_{i=1}^{I} \eta_i'(S)N_i + R^2 Q_0 \, e^{-Rt} \\
&= -\sum_{i=1}^{I} \left(\eta_i(S) - R \right)^2 N_i - \dot{S} \sum_{i=1}^{I} \eta_i'(S)N_i + R^2 Q_0 \, e^{-Rt}.
\end{aligned}
$$

As in the case discussed in Section 6.1.2, we can multiply by $\mathrm{sgn}_+(\dot{S})$ and arrive at a Lyapunov functional

$$
\begin{aligned}
\frac{d}{dt} \left(\dot{S}(t) \right)_+ &\leq -\left(\dot{S}(t) \right)_+ \sum_{i=1}^{I} \eta_i'(S)N_i(t) + R^2 |Q_0| \, e^{-Rt} \\
&\leq -\alpha \underline{M} \left(\dot{S}(t) \right)_+ + R^2 |Q_0| \, e^{-Rt}
\end{aligned}
$$

because $\eta_i' \geq \alpha > 0$ (by assumption (1.28)), and using step 2. From this inequality and the fact that $\dot{S}(t)$ is bounded, we deduce that

$$\alpha \underline{M} \int_{t=0}^{\infty} \left(\dot{S}(t) \right)_+ dt \quad \text{is finite.}$$

This proves also (still because $S(t)$ is bounded by the first step) that $S(t)$ is of Bounded Variation on $\mathbb{R}^+$, i.e., $\int_{t=0}^{\infty} |\dot{S}(t)| dt$ is finite, see Section 6.5.1 for details. Therefore it has a limit as $t \to \infty$.

Fourth step. Identifying the limits. We prove that

$$\lim_{t\to\infty} S(t) = S^*.$$

Indeed, otherwise we would have (i) if $\lim_{t\to\infty} S(t) > S^*$, then $\lim_{t\to\infty} \eta_{i_0}(S(t)) > R$ and then $N_{i_0}(t)$ has an exponential growth, which contradicts the bound proved in step 1, (ii) if $\lim_{t\to\infty} S(t) < S^*$, then $\lim_{t\to\infty} \eta_i(S(t)) < R$ for all $1 \leq i \leq I$ and $N_i(t)$ vanishes exponentially which contradicts step 2.

Then, we conclude the lemma because $\lim_{t\to\infty} \eta_i(S(t)) < R$ for all $i \neq i_0$ and thus $N_i(t)$ vanishes exponentially. Therefore, using step 1, $N_{i_0}(t)$ tends to $S^0 - S^*$. □

We refer to [211] for other results in this direction and a general view on chemostat. As a consequence of Lemma 1.1, there cannot be coexistence of several micro-organisms in such a model, in opposition to observations (in nature or sometimes laboratories); we refer to [160] for a variant which allows the coexistence of several micro-organisms.

1.6.2   Continuous model of a chemostat

We now present a natural extension of the previous differential system to a continuous uptake ability. We have taken this example from [83] and from Chapter 2. It is obtained formally by saying that the index i becomes a continuous variable denoted by x below, and this implies

$$\begin{cases} \dot{S}(t) = R(S_0 - S) - \int_{x>0} \eta(x, S)n(t, x)dx, \quad t \geq 0, \ x > 0, \\[2mm] \frac{\partial}{\partial t}n(t, x) = n(t, x)\big(\eta(x, S) - R\big), \\[2mm] S(0) = S^0 > 0, \quad n(t = 0, x) = n^0(x) > 0, \ n^0 \in L^1 \cap L^\infty(\mathbb{R}^+). \end{cases} \quad (1.30)$$

To make this model more interesting one can add mutations as we do in Chapter 2 or [45, 83].

One can easily check that the proof of Lemma 1.1 extends to this physiologically structured population model,

Theorem 1.2. *Assume*

$$\eta(x, \cdot) \text{ is increasing,} \qquad \eta(x, S_0) > R, \ \forall x > 0,$$

and denoting the inverse with respect to S by η^{-1}, assume that

$$\exists x_0 > 0, \quad S^* := \eta^{-1}(x_0, R) < \eta^{-1}(x, R) \quad \forall x \neq x_0.$$

Then, one has, as $t \to \infty$,

$$N(t,x) \to (S_0 - S^*)\delta(x = x_0), \qquad S(t) \to S^*.$$

Such a population which has a single trait is called monomorphic. We refer to Chapter 2, Sections 2.1 and 2.2 for proofs of a similar selection principle.

1.7 Phytoplankton (light structure)

A nonlinear model has been proposed to explain why phytoplankton does not sink while it should, being denser than water. In this section, we present the model and give an introductory analysis. More complete mathematical results can be found in [144].

A good survey of the subject can be found in [117, 116] together with numerical results. The word phytoplankton covers several species of photosynthesizing microscopic (2 to 200 μm) organisms (by opposition to zooplankton, a predator for phytoplankton). For that reason it represents the first stage of an alimentary chain and an important cause of carbon absorption from atmosphere. These species inhabit the upper layer of oceans or lakes (50 to 100 m) where light is available, but they are 2 to 5 percent denser than water. The reasons why phytoplankton can sustain under these circumstances are various. Some species are able to swim by means of flagella, some other species are endowed with gas vacuoles (in the upper layers light creates an increase of weight and of density, but after sinking, food restriction makes them lighter). Another explanation for up-welling is mixing effects for the lake or ocean which helps minerals come from deep water to the surface and also creates a 'turbulent' suspension of the population that can reproduce before ultimately sinking. More about this wide subject can be found in [203].

The full model is a partial differential equation but one can consider an isolated water column of still water, and we denote by z the vertical coordinate directed downward. Water surface is at $\{z = 0\}$ and thus the model is posed for $z \geq 0$ (z axis downward), $t \geq 0$,

$$\begin{cases} \frac{\partial}{\partial t}n(t,z) + v_p\frac{\partial}{\partial z}n(t,z) - \kappa\frac{\partial^2}{\partial z^2}n(t,z) = f(z, S(t,z,[n]))\, n(t,z), \\[2mm] \kappa\frac{\partial}{\partial z}n(t,0) + v_p n(t,0) = 0, \\[2mm] n(t,z) \to 0 \qquad \text{as } z \to \infty. \end{cases} \tag{1.31}$$

As usual $n(t,z)$ denotes the population density at time t and depth z. The diffusion term takes into account the random motion due to fluctuating water motion, $v_p > 0$ is the vertical transport velocity of gravitational sinking. The birth/death rate $f(z, S(t,z,[n]))$ depends on the local light available (and thus it is a decreasing function of z), it is positive for small values of z and negative for large z. But it also takes into account the 'shading' effect on a layer by the phytoplankton above it $S(t, z, [n]) = \int_0^z n(t, \bar{z})d\bar{z}$. An example is

$$
\begin{cases}
f\big(z, S(t, z, [n])\big) = B\big(z\,[1 + \sigma \int_0^z n(t, \bar{z})d\bar{z}]\big), \\
B(0) > 0, \quad B(\infty) < 0, \quad B' < 0.
\end{cases}
\tag{1.32}
$$

A complete analysis of the model is given in [144]. A condition on the data is needed for existence of a (unique) stationary solution. When this condition is fulfilled, the solution $n(t, z)$ converges as $t \to \infty$ to this steady state. When it is not fulfilled, it means that the growth zone ($B > 0$) is too small, and whatever is the initial data, the solution vanishes for large times.

Here, we restrict ourselves to the so-called layer model of [117, 116]. It consists in a simple case where B has a (single) discontinuity. Then there are two zones, a euphotic zone $z < H$ and an aphotic zone $z > H$, and two real numbers B^+, B^-, and we take

$$
B(q) = \begin{cases}
B^+ > 0 & \text{for} \quad q < H_0, \\
B^- < 0 & \text{for} \quad q > H_0,
\end{cases}
\tag{1.33}
$$

and this leads to a discontinuity at the layer $z = H$ implicitly defined by the equation

$$
H_0 = H\big(1 + \sigma \int_0^H n(t, \bar{z})d\bar{z}\big).
\tag{1.34}
$$

It admits a solution because, n being nonnegative, the right-hand side is an increasing function of H from 0 to ∞.

For the *layer model*, the stationary problem can be solved analytically.

Theorem 1.3 (Layer model). *For $\kappa > \frac{v_p^2}{4B_+}$, there is a unique stationary nonnegative solution $n(x) \in C^1(\mathbb{R}^+)$ to (1.31)–(1.33) iff $H_0 > H$ for $H(\kappa, v_p, B^\pm)$ a positive parameter given in the proof below.*

Remark 1.2. The general condition which replaces the above two conditions in the layer case is simply that the solution to

$$
\begin{cases}
v_p \frac{\partial}{\partial z} n(z) - \kappa \frac{\partial^2}{\partial z^2} n(z) = f(z, 0)\, n(z), \\
\kappa \frac{\partial}{\partial z} n(0) + v_p n(0) = 0 \quad n(0) = 1,
\end{cases}
\tag{1.35}
$$

does not remain positive for all $z > 0$.

Proof. First step. Fixed discontinuity. We first prove that there is a unique discontinuity parameter $H \geq 0$ for which there is a positive solution n_H to equation

$$
\begin{cases}
\kappa n_H(z)'' - v_p n_H(z)' + B(z)\, n_H(z) = 0, & z > 0, \\[2mm]
\kappa n'_H(0) - v_p n_H(0) = 0, & n_H(\infty) = 0, \\[2mm]
B(z) = B^+ \; for \; z < H, \qquad B(z) = B^- \; for \; z > H.
\end{cases}
\tag{1.36}
$$

This is an eigenvalue problem and thus the solution is defined up to a multiplicative constant. Hence for $z > H$, we may choose $n_H(z) = e^{-\gamma z}$, with $-\gamma$ the negative root of the characteristic equation $\kappa X^2 - v_p X + B^- = 0$, i.e., $\gamma = \frac{1}{2\kappa}(\sqrt{v_p^2 + 4\kappa|B^-|} - v_p)$.

For $0 < z < H$, the solution space is the two-dimensional vector space $n_H(z) = Ae^{\alpha z} + Be^{\beta z}$, $A \in \mathbb{C}$, $B \in \mathbb{C}$, with α and β the solutions to the characteristic equation $\kappa X^2 - v_p X + B^+ = 0$ which, under the stated condition on κ are given by,

$$
\alpha = \frac{1}{2\kappa}(v_p - \imath\sqrt{4\kappa B^+ - v_p^2}), \qquad \beta = \frac{1}{2\kappa}(v_p + \imath\sqrt{4\kappa B^+ - v_p^2}).
$$

Then, we have three conditions on A and B coming from the boundary condition on the surface level $z = 0$, the continuity and the continuity of derivatives at $z = H$,

$$
A(v_p - \kappa\alpha) + B(v_p - \kappa\beta)B = 0,
$$
$$
Ae^{\alpha H} + Be^{\beta H} = e^{-\gamma H},
$$
$$
\alpha Ae^{\alpha H} + \beta Be^{\beta H} = -\gamma e^{-\gamma H}.
$$

The existence of a unique nontrivial solution amounts to writing

$$
0 = \det \begin{pmatrix} v_p - \kappa\alpha & v_p - \kappa\beta \\ (\gamma + \alpha)e^{\alpha H} & (\gamma + \beta)e^{\beta H} \end{pmatrix}
$$

which is reduced to the simple equation on H,

$$
(v_p - \kappa\alpha)(\gamma + \beta)e^{\beta H} - (v_p - \kappa\beta)(\gamma + \alpha)e^{\alpha H} = 0.
$$

Therefore, we find the condition that determines H,

$$
e^{(\beta - \alpha)H} = \frac{\gamma + \alpha}{\gamma + \beta} \frac{v_p - \kappa\beta}{v_p - \kappa\alpha},
$$

or, after introducing the notation

$$
\Delta^{\pm} = \sqrt{4\kappa|B^{\pm}| \mp v_p^2},
$$

we end up with

$$\exp\left(\imath\, H\, \frac{\Delta^+}{\kappa}\right) = \frac{\Delta^- - \imath\Delta^+}{\Delta^- + \imath\Delta^+}\, \frac{v_p + \imath\Delta^+}{v_p - \imath\Delta^+}. \tag{1.37}$$

Because the complex number on the right-hand side has modulus 1, there is a single root H with $H\,\frac{\Delta^+}{\kappa} \in (0, 2\pi)$. Also we claim that its imaginary part is nonnegative and thus

$$H\, \frac{\Delta^+}{\kappa} \in (0, \pi). \tag{1.38}$$

Indeed, the sign of its imaginary part is also that of

$$[(\Delta^- - \imath\Delta^+)(v_p + \imath\Delta^+)]^2 = [v_p\Delta^- + (\Delta^+)^2 + \imath(\Delta^- - v_p)\Delta^+]^2$$

$$= \text{ real part } + 2\imath(v_p\Delta^- + (\Delta^+)^2)(\Delta^- - v_p)\Delta^+.$$

And we obtain the positivity of this imaginary part because $\Delta^- > v_p$.

We now claim that this is the only possible choice of $H > 0$ which allows us to obtain a positive solution n. Indeed, because $B = \bar{A}$, $\beta = \bar{\alpha}$, we have for $0 \leq z \leq H$ and by continuity at $z = H$,

$$n_H(z) = 2\mathcal{R}e\left(Ae^{\alpha z}\right) = e^{-\gamma H}\,\mathcal{R}e\left(e^{\alpha(z-H)}\right) = e^{-\gamma H}\, e^{-\frac{v_p}{2\kappa}(z-H)}\, \cos\!\left(\frac{\Delta^+}{2\kappa}(z - H)\right).$$

This quantity is positive if and only if $\cos(\frac{\Delta^+}{2\kappa}H) > 0$, which holds true thanks to the above condition (1.38).

Second step. Nonlinear problem. Thanks to the first step there is a single possibility for having a positive stationary solution. Namely, we have to find a positive constant C such that $n = Cn_H$, recalling that n_H denotes the solution built in the first step. This is equivalent to saying that

$$H_0 = H[1 + C\sigma \int_0^H n_H(z)dz],$$

an equation which has a solution $C > 0$ if and only if $H_0 > H$. $\square$

Chapter 2

Adaptive dynamics; an asymptotic point of view

So far, we have given examples of physiologically structured populations, i.e., structured by a parameter describing a biological, physiological or ecological characteristic of the individuals. Further examples are also described in Chapter 4 below. When this characteristic is inherent to the individual, i.e., it is fixed at the very beginning of its life, we refer to it as a *trait*; we prefer to avoid calling it a phenotype. The theory which focusses on phenotypic evolution driven by *small mutations* in replication, while ignoring both sex and genes, is known by the name *Adaptive dynamics* and is part of *Evolution theory*, see [172, 115, 80, 81, 169, 45] and the references given therein. A general mathematical treatment of the general subject of selection vs mutation can also be found in [43] (in particular population geneticists might prefer the assumption that mutations are rare rather than small).

The two main ingredients in this theory are (i) the selection principle which favorizes the population with best adapted trait, and (ii) mutations which allow off-springs to have slightly different traits than their mother. The combination of the two effects is studied by adaptive dynamics. This turns out to be an extremely intricate theory on which several possible mathematical approaches are possible. One of the reasons is that it is merely impossible to consider this problem without introducing small parameters (mutations can be small or rare for instance, population should be large in any case but relative death rates can vary). Therefore *adaptive evolution* can be studied with various mathematical tools. Evolutionary game theory is a standard point of view, see for instance [135, 136, 134] after it was introduced by J. Maynard Smith (see [169]). Probability theory is also natural because fluctuations are important at the individual level and individual centered models are naturally stochastic. Departing from an individual centered stochastic dynamics, several possible limits are possible as the population becomes large ([101]). Here we look at models which are in one special class of those limits.

In this chapter, we give a first and very elementary point of view based on structuring an ODE and including mutations. Our main goal is to apply the corresponding asymptotic theory and show how the concept of monomorphic population arises naturally in the limit of small mutations over a long time compared to one generation length. Such rescaling has also been used for the spread of genetic traits, in a probabilistic framework (see [110] and the references therein).

Our presentation in this chapter follows the lines of [83] but uses a simpler framework for the population model. We begin with several simple examples of physiologically structured population (without mutations) where a selection principle can be proved. Then we introduce the mutations and this raises the question of finding the appropriate scales. Here, and this is one the possible scales, we assume that mutations are frequent but have a very small effect on the trait. This allows us to state an asymptotic problem. We show that it leads to the selection of a single trait (monomorphic population) but the dynamics of this trait is far from obvious. A differential equation is not enough to describe it and we introduce, following [83], a Hamilton–Jacobi equation. The possible asymptotic behaviors of this trait (monomorphic or dimorphic) are obtained through numerical simulations: continuous evolution, jump, branching from a monomorphic to a dimorphic population or junction from dimorphic to monomorphic situations.

Our motivation for the model we use in this chapter comes from the chemostat equation (1.30). Consider that the time scale for food consumption is faster than the time scale of birth, then one naturally ends up with an algebraic equation for the substrate that is the system

$$
\begin{cases}
S(t) = S_0 - \int \eta\big(x, S(t)\big)n(t,x)dx, & t \geq 0,\ x > 0 \\[2mm]
\frac{\partial}{\partial t}n(t,x) = n(t,x)\big(\eta(x,S) - R\big), & \\[2mm]
n(t = 0, x) = n^0(x) > 0,\ n^0 \in L^1 \cap L^\infty(\mathbb{R}^+).
\end{cases}
\tag{2.1}
$$

When η is linear in S, we arrive at the more explicit formula

$$
S(t) = Q_b\Big(\int \eta(x)n(t,x)dx\Big)
$$

and this is the form we take in this chapter as a compromise between generality and simplicity.

2.1 Structured population and selection principle: a simple case

The first question we consider here is how to give a mathematical description of the process in which some specific trait is selected in a given environment. It is the best adapted trait in terms of using resources and that trait is called an *Evolutionary Stable Strategy* (ESS in short). The origin of this denomination ([170]) comes from evolution theory; no mutant with a different trait can invade a population with the trait corresponding to an ESS. An example of the selection principle has already been mentioned, in the case of the chemostat, see Lemma 1.2. Here we give an easy example that can be treated by explicit computations. In Section 2.2 we give a more general framework.

To begin with a simple example, we look at a variant of Verhulst's logistic equation (see Section 1.1) which we structure with a trait $x \in \mathbb{R}$ and we illustrate the selection principle on this very simple example. We consider that the reproduction rate depends on the trait, i.e., $b = b(x) > 0$, and that the death rate is proportional to the total population number (this is what happens in logistic growth, see Section 1.1). We arrive at

$$
\begin{cases}
\frac{d}{dt}n(t,x) & = b(x)n(t,x) - \varrho(t)n(t,x), \\[2mm]
\varrho(t) & = \int_{\mathbb{R}} n(t,x)dx, \\[2mm]
n(t=0) & = n^0(x) > 0 \text{ for } x \in]x_m, x_M[, \qquad n^0(x) = 0 \text{ for } x \notin]x_m, x_M[.
\end{cases}
\tag{2.2}
$$

In this situation one can guess that the best adapted trait (highest reproduction rate here) will be selected. Indeed we have the following selection principle.

Theorem 2.1. *Assume that $b \in C(\mathbb{R})$, $b(\cdot) \geq \underline{b} > 0$, $n^0 > 0$ and*

$$
b(\bar{x}) = \max_{x \in [x_m, x_M]} b(x) \quad \text{is attained for a single } \bar{x} \in]x_m, x_M[.
$$

Then, the solution to (2.2) satisfies

$$
\varrho(t) \to \bar{\varrho} = b(\bar{x}), \qquad n(t,x) \rightharpoonup b(\bar{x})\delta(x - \bar{x}) \qquad \text{as } t \to \infty.
\tag{2.3}
$$

Notice that the equation (2.2) admits many steady states, namely $n(x) = \varrho\,\delta(x = y)$, $\varrho = b(y)$ for any y, Theorem 2.1 selects the stable trait, the ESS. One should understand it as the trait that realizes

$$
\max_x[b(x) - \bar{\varrho}] = b(\bar{x}) - \bar{\varrho} = 0.
$$

The weak limit in (2.3) also indicates that the natural setting for structured population models should differ from that for differential equations because functional spaces (measures here) appear to play a role.

Proof. We give a proof that relies on a simple computation, another proof is possible based on more abstract arguments and that apply to more general equations (see Section 2.2). We define

$$
N(t,x) = n(t,x)e^{\int_0^t \varrho(s)ds}.
\tag{2.4}
$$

This satisfies

$$\frac{dN(t,x)}{dt} = b(x)N(t,x),$$

and thus $N(t,x) = n^0(x)e^{b(x)t}$. We deduce from (2.4) that

$$\frac{d}{dt}e^{\int_0^t \varrho(s)ds} = \varrho(t)e^{\int_0^t \varrho(s)ds} = \int_{\mathbb{R}} N(t,x)dx = \int_{\mathbb{R}} n^0(x)e^{b(x)t}dx.$$

Therefore

$$e^{\int_0^t \varrho(s)ds} = \int_{\mathbb{R}} \frac{n^0(x)}{b(x)}e^{b(x)t}dx + K, \qquad K = 1 - \int_{\mathbb{R}} \frac{n^0(x)}{b(x)}dx,$$

$$\int_0^t \varrho(s)ds = \ln\left(\int_{\mathbb{R}} \frac{n^0(x)}{b(x)}e^{b(x)t}dx + K\right), \tag{2.5}$$

$$\varrho(t) = \int_{\mathbb{R}} n^0(x)e^{b(x)t}dx / \left[\int_{\mathbb{R}} \frac{n^0(x)}{b(x)}e^{b(x)t}dx + K\right],$$

and we notice that the constant K may be negative but the denominator is larger than 1. This is a Laplace type formula and we can analyze it as follows. As $t \to \infty$, we have

$$\varrho(t) \le b(\bar{x}) \int_{\mathbb{R}} \frac{n^0(x)}{b(x)}e^{b(x)t}dx / \left[\int_{\mathbb{R}} \frac{n^0(x)}{b(x)}e^{b(x)t}dx + K\right] \to b(\bar{x}),$$

because $\int_{\mathbb{R}} \frac{n^0(x)}{b(x)}e^{b(x)t}dx \to \infty$. For the reverse inequality, we fix an $\varepsilon > 0$ and define the set

$$I(\varepsilon) = \{x;\ b(x) \ge b(\bar{x}) - \varepsilon\}.$$

Then, we have

$$\varrho(t) \ge \int_{I(\varepsilon)} n^0(x)e^{b(x)t}dx / \left[\int_{\mathbb{R}} \frac{n^0(x)}{b(x)}e^{b(x)t}dx + K\right]$$

$$\ge (b(\bar{x}) - \varepsilon) \int_{I(\varepsilon)} \frac{n^0(x)}{b(x)}e^{b(x)t}dx / \left[\int_{\mathbb{R}} \frac{n^0(x)}{b(x)}e^{b(x)t}dx + K\right]$$

$$\ge \frac{b(\bar{x}) - \varepsilon}{A_\varepsilon(t)},$$

with

$$A_\varepsilon(t) = \frac{K}{\int_{I(\varepsilon)} \frac{n^0(x)}{b(x)}e^{b(x)t}dx} + \frac{\int_{\mathbb{R}} \frac{n^0(x)}{b(x)}e^{b(x)t}dx}{\int_{I(\varepsilon)} \frac{n^0(x)}{b(x)}e^{b(x)t}dx} \xrightarrow{t\to\infty} 1.$$

Therefore, we have indeed proved the first statement, namely that

$$\varrho(t) \to b(\bar{x}) \quad as \quad t \to \infty.$$

Finally, from (2.4) and the expression of $N(t, x)$, we deduce

$$n(t, x) = n^0(x) e^{b(x)t} e^{-\int_0^t \varrho(s)ds},$$

and the claim (2.3) on $n(t, x)$ follows from the limit of $\varrho(t)$ combined with the same arguments for this Laplace formula. Indeed, for $x \neq \bar{x}$, we have $b(x) < b(\bar{x})$ and thus

$$n(t, x) = n^0(x) e^{-\int_0^t (\varrho(s)-b(x))ds} \to 0, \qquad t \to \infty.$$

This proves the result. $\qquad\qquad\qquad\qquad\qquad\qquad\qquad\qquad\qquad\qquad\qquad\qquad$ □

2.2 Structured population and selection principle: an extension

We now prove a selection principle in a more general setting which will be used later on and does not allow for explicit formulas as in the previous section, namely

$$
\begin{cases}
\frac{d}{dt} n(t, x) &= \left[b(x) Q_b(\varrho(t)) - d(x) Q_d(\varrho(t)) \right] n(t, x), \quad x \in \mathbb{R}, \\[2mm]
\varrho(t) &= \int_{\mathbb{R}} n(t, x) dx, \\[2mm]
n(t = 0) &= n^0(x) > 0.
\end{cases}
\tag{2.6}
$$

The term $b(x)Q_b(\varrho(t)) - d(x)Q_d(\varrho(t))$ can be interpreted as the fitness of individuals with the trait x being given the environment created by the total population. Here the notation is the same as in the previous section and we assume that $b, d \in C(\mathbb{R})$, Q_b, $Q_d \in C^1(\mathbb{R}^+)$ and there are b_m, b_M, d_M and d_m such that

$$b_M \geq b(\cdot) \geq b_m > 0, \qquad d_M \geq d(\cdot) \geq d_m > 0, \tag{2.7}$$

$$
\exists\, 0 < \varrho_m < \varrho_M \quad \text{such that} \quad
\begin{cases}
\max_{x \in \mathbb{R}} \left[b(x) Q_b(\varrho_M) - d(x) Q_d(\varrho_M) \right] < 0, \\[2mm]
\min_{x \in \mathbb{R}} \left[b(x) Q_b(\varrho_m) - d(x) Q_d(\varrho_m) \right] > 0,
\end{cases}
\tag{2.8}
$$

$$
\begin{cases}
\text{there is a single pair } (\bar{x}, \bar{\varrho}) \in \mathbb{R} \times [\varrho_m, \varrho_M] \quad \text{such that} \\[2mm]
b(\bar{x}) Q_b(\bar{\varrho}) - d(\bar{x}) Q_d(\bar{\varrho}) = 0 = \max_{x \in \mathbb{R}} \left[b(x) Q_b(\bar{\varrho}) - d(x) Q_d(\bar{\varrho}) \right].
\end{cases}
\tag{2.9}
$$

Notice that there exists a unique $\bar{\varrho}$ that satisfies the equality

$$0 = \max_{x \in \mathbb{R}} \left[b(x) Q_b(\bar{\varrho}) - d(x) Q_d(\bar{\varrho}) \right]$$

in (2.9). It is indeed a consequence of (2.8) and of the additional assumption

$$Q_b'(\varrho) < 0, \qquad Q_d'(\varrho) > 0. \tag{2.10}$$

In (2.9) we assume additionally that the maximum is attained by a single point $\bar{x}$; this is unessential but simplifies the statement below. Finally, we also need that there is $R > 0$ such that for ϱ close enough to $\bar{\varrho}$, then

$$\beta_R := \max_{|x| \geq R} \left[b(x)Q_b(\varrho) - d(x)Q_d(\varrho) \right] < 0. \tag{2.11}$$

Again the selection principle applies in this case and gives

Theorem 2.2. *With assumptions (2.7)–(2.11), and assuming that $n^0(x) > 0$ and $\varrho_m \leq \varrho(t = 0) \leq \varrho_M$, the solution to (2.6) satisfies*

$$\varrho_m \leq \varrho(t) \leq \varrho_M, \qquad \forall t \geq 0, \tag{2.12}$$

$$\varrho(t) \to \bar{\varrho}, \quad n(t, x) \to \bar{\varrho}\delta(x = \bar{x}), \qquad as \ t \to \infty. \tag{2.13}$$

In other words, we recover that the ESS, the unbeatable strategy, is characterized by (2.9) as in the previous section.

Proof. First step. A priori estimate. We first prove (2.12) using assumption (2.8). Integrating the equation (2.6) in x, we obtain

$$\min_{x \in \mathbb{R}} \left[b(x)Q_b(\varrho(t)) - d(x)Q_d(\varrho(t)) \right] \varrho(t) \leq \frac{d}{dt} \varrho(t)$$

$$\leq \max_{x \in \mathbb{R}} \left[b(x)Q_b(\varrho(t)) - d(x)Q_d(\varrho(t)) \right] \varrho(t).$$

Therefore, if $\varrho(t)$ approaches ϱ_m, it is increasing and thus we have for all times $\varrho_m < \varrho(t)$, and similarly if $\varrho(t)$ approaches ϱ_M it is decreasing and thus we have for all times $\varrho(t) < \varrho_M$.

Second step. A Lyapunov functional. We consider a function $P(r)$ satisfying

$$rP'(r) + P(r) = Q(r) := \frac{Q_d(r)}{Q_b(r)}, \tag{2.14}$$

a differential equation on the interval $[\varrho_m, \varrho_M]$ with several solutions given by $P_0(r) + \frac{\lambda}{r}$ (but this family is in fact reduced to a single function as far as we are interested in the expression below) .

Then we compute

$$\frac{d}{dt} \int [\frac{b(x)}{d(x)} - P(r)] n(t, x)dx$$

$$= \int [\frac{b(x)}{d(x)} - P(r) - \varrho(t)P'(\varrho(t))] \left[b(x)Q_b(\varrho(t)) - d(x)Q_d(\varrho(t)) \right] n(t, x)dx$$

$$= \int \frac{1}{d(x)Q_b(\varrho(t))} \left[b(x)Q_b(\varrho(t)) - d(x)Q_d(\varrho(t)) \right]^2 n(t, x)dx.$$

As a consequence the bounded quantities $\int \left[\frac{b(x)}{d(x)} - P(r)\right] n(t,x) dx$ are increasing and thus converge as $t \to \infty$,

$$\int \left[\frac{b(x)}{d(x)} - P(r)\right] n(t,x) dx \xrightarrow[t\to\infty]{} L \in \mathbb{R}, \qquad (2.15)$$

$$\int_0^\infty \int_\mathbb{R} \left[b(x)Q_b(\varrho(t)) - d(x)Q_d(\varrho(t))\right]^2 n(t,x) dx dt < \infty. \qquad (2.16)$$

Third step. The limit. We now derive a limit for another quantity (we recall the definition of $Q(r)$ in (2.14))

$$\frac{d}{dt} \int \left[\frac{b(x)}{d(x)} - Q(\varrho(t))\right]^2 n(t,x) dx$$

$$= \int \left[\frac{b(x)}{d(x)} - Q(\varrho(t))\right]^2 \left[b(x)Q_b(\varrho(t)) - d(x)Q_d(\varrho(t))\right] n(t,x) dx$$

$$- 2Q'((\varrho(t)) \int \left[\frac{b(x)}{d(x)} - Q(\varrho(t))\right] n(t,x) dx$$

$$\times \int \left[b(x)Q_b(\varrho(t)) - d(x)Q_d(\varrho(t))\right] n(t,x) dx.$$

Therefore, using the L^∞ and non-extinction bound to upper bound the first term in the right-hand side, and Cauchy-Schwarz inequality for the second term, we also have

$$\left|\frac{d}{dt} \int \left[\frac{b(x)}{d(x)} - Q(\varrho(t))\right]^2 n(t,x) dx\right| \leq C \int \left[b(x)Q_b(\varrho(t)) - d(x)Q_d(\varrho(t))\right]^2 n(t,x) dx.$$

As a conclusion of this computation and step (ii), we deduce that

$$\frac{d}{dt} \int \left[\frac{b(x)}{d(x)} - Q(r)\right]^2 n(t,x) dx \in L^1(\mathbb{R}^+).$$

This is stronger than a mere BV bound which is enough to ensure that it has a limit, , see Section 6.5.1 for details. Thanks to its integrability (second conclusion of step (ii)), we conclude

$$\int \left[\frac{b(x)}{d(x)} - Q(\varrho(t))\right]^2 n(t,x) dx \xrightarrow[t\to\infty]{} 0.$$

And again the use of Cauchy-Schwarz inequality results in

$$\int \left|\frac{b(x)}{d(x)} - Q(\varrho(t))\right| n(t,x) dx \xrightarrow[t\to\infty]{} 0. \qquad (2.17)$$

Combined with the first conclusion of step (ii), we also arrive at

$$\int \left[\frac{b(x)}{d(x)} - P(\varrho(t))\right] n(t,x) dx = \int \left[\frac{b(x)}{d(x)} - Q(\varrho(t)) + [Q-P](\varrho(t))\right] n(t,x) dx \to L,$$

and thus $[Q - P](\varrho(t))\,\varrho(t)$ has a limit and finally

$$\varrho(t) \xrightarrow[t \to \infty]{} \varrho^*. \tag{2.18}$$

Indeed, on the one hand $Q - P$ is not locally constant because

$$r(P - Q)' + P - Q = -rQ' < 0,$$

and on the other hand $\varrho(t)$ is Lipschitz continuous (to see this, integrate in x the equation on n). Notice at this point that we do not use the full assumption (2.10) but only the consequence that Q' does not vanish.

Fourth step. Identifying $\varrho^* = \bar{\varrho}$. Indeed, if $\varrho^* > \bar{\varrho}$, then for t large enough we have

$$\max_{x \in \mathbb{R}}[b(x)Q_b(\varrho(t)) - d(x)Q_d(\varrho(t))] < 0,$$

and the differential inequality of step (i) (upper bound) implies that there is extinction which is impossible since $\varrho(t) \geq \rho_m$. Also, if $\varrho^* < \bar{\varrho}$, then for t large enough we have

$$\max_{x \in \mathbb{R}}[b(x)Q_b(\varrho(t)) - d(x)Q_d(\varrho(t))] > 0,$$

and for those x's where $[b(x)Q_b(\varrho(t)) - d(x)Q_d(\varrho(t))] > 0$ we have exponential growth, i.e., blow-up for $\varrho(t)$ which is again a contradiction.

Fifth step. Weak limit for $n(t, x)$. From (2.11), we have for t large enough

$$\frac{d}{dt}\int_{|x|>R} n(t,x)dx \leq \beta_R \int_{|x|>R} n(t,x).$$

Therefore $\sup_{t>0}\int_{|x|>R} n(t,x)dx \to 0$ as $R \to \infty$ (because this is true initially). This proves that the family $\big(n(t,x)\big)_{t>0}$ is compact in the weak sense of measures. Therefore there are subsequences that converge weakly to measures $n^*(x)$ and $\bar{\varrho} = \int n^*(x)dx$ (we denote integration with measures as Lebesgue integrals to avoid specific notation).

Sixth step. Conclusion. From (2.17)–(2.18), we know that $n^*(x)$ should be concentrated on the set of x's such that $b(x)Q_b(\bar{\varrho}) - d(x)Q_d(\bar{\varrho}) = 0$. With assumption (2.9), this point is unique and thus $n^*(x) = \bar{\varrho}\delta(x = \bar{x})$. Therefore the full family $n(t,x)$ converges (all subsequences have the same limit!) and the result is proved. $\square$

2.3   Structured population: cannibalism

Models such as (2.6) are still not as general as one might wish. The simplified model of a chemostat in (2.1) provides us with an example. Cannibalism is another example of physiologically structured population where not only the total population counts. We borrow the model from [81] and references therein.

Here, the trait $x \geq 0$ under consideration is the degree of cannibalism, and $0 < \alpha \leq 1$ denotes the efficiency in offspring production from intraspecific predation, the (irrealistic) model is then written, for $t \geq 0$, $x \geq 0$, as

$$
\begin{cases}
\frac{\partial}{\partial t} n(t,x) = r n(t,x) - \int_0^\infty y n(t,y) dy \, n(t,x) + \alpha \, x \, n(t,x) \int_0^\infty n(t,y) dy, \\
n(t = 0, x) = n^0(x) > 0.
\end{cases}
$$

(2.19)

The parameter r is just the growth rate in the absence of cannibalism and one can see that cannibalism acts as a negative feedback. The growth rate of the population is indeed lower than the logistic rate r since

$$
\frac{d}{dt} \int_0^\infty n(t,x) = \int_0^\infty n(t,x) \big[r - (1-\alpha) \int_0^\infty x \, n(t,x) \big] \leq r \int_0^\infty n(t,x). \quad (2.20)
$$

In fact, one can go further and prove that such a population will go extinct, a behavior rather different from that of the previous sections and that can be interpreted as an ESS located at $x = \infty$.

Theorem 2.3. *Assume that the initial data in* (2.19) *decays in x so that*

$$
\int_0^\infty n^0(x) e^{x^2} \, dx < \infty \quad \text{and} \quad \int_0^\infty x n^0(x) dx \geq \frac{r}{1-\alpha},
$$

then there is a global positive solution to (2.19) *and we have the time decay*

$$
\int_0^\infty n(t,x) dx \quad \text{decays to } 0 \text{ as } t \to \infty.
$$

Remark 2.1. 1. Since, as we see in the proof, $\int_0^\infty x n(t,x) dx$ remains positive, we conclude that the population is concentrated on a very large trait for large times as announced above.

2. The assumption $\int_0^\infty x n^0(x) dx \geq \frac{r}{1-\alpha}$ is not necessary (see the exercise below).

3. Of course one can doubt the possibility to undergo unlimited growth rates x and this leads to pose the same problem with $0 \leq x \leq A$. Then the method is similar and one can prove that the total population converges to $\rho \delta(x = A)$, in other words for this model the ESS is $x = A$ (see exercise below).

Proof. We set $m_k(t) = \int_0^\infty x^k n(t,x) dx$.

First step. Existence. From (2.20), we derive the a priori estimate $m_0(t) \leq m_0(0) e^{rt}$. Therefore we have the upper bound

$$
n(t,x) = n^0(x) e^{rt + \alpha x \int_0^t m_0(s) - \int_0^t m_1(s)} \leq n^0(x) e^{rt + \alpha x m_0(0) e^{rt}/r}.
$$

This proves that, after using the Cauchy–Lipschitz Theorem, the solutions do not blow-up in finite time and thus they are global. In particular, thanks to our assumption on the initial growth in x, the $m_k(t)$ are finite but there is no extinction in finite time.

Second step. A lower bound on m_1. We have (using Cauchy–Schwarz inequality) some kind of Lyapunov functional

$$\frac{d}{dt}m_1(t) = m_1(t)(r - m_1(t)) + \alpha m_2(t)m_0(t) \geq m_1(t)\big(r - m_1(t)(1 - \alpha)\big). \quad (2.21)$$

Therefore we find that for all times

$$m_1(t) \geq \frac{r}{1 - \alpha}. \quad (2.22)$$

As a consequence of (2.20), we deduce that $m_0(t)$ decays with time to a limit $L_0 \geq 0$.

Third step. Limits for the $m_k(t)$. We further write

$$\frac{d}{dt}\frac{m_{k+1}(t)}{m_k(t)} = \alpha \frac{m_0(t)}{m_k^2(t)}\big(m_{k+2}(t)m_k(t) - m_{k+1}{}^2(t)\big) \geq 0.$$

Therefore the above ratio has a limit and thus $m_1(t) \to L_1$ as $t \to \infty$.

If $L_1 = \infty$, obviously from (2.20) we deduce $L_0 = 0$ and the theorem is proved. We now assume, by opposition, that $L_1 < +\infty$, $L_0 > 0$ and thus $m_2(t)$ has a limit L_2.

Fourth step. A contradiction. From the inequality (2.21), we deduce that $L_1 = \frac{r}{1-\alpha}$ and this is a contradiction with the exact solution $n(t, x)$ given in step (i). □

Exercise. In the case $0 \leq x \leq A$, prove that $L_0 \neq 0$ and that $n \to \rho\delta(x = A)$. To do that check that in the proof above

$$L_2 L_0 = L_1^2.$$

Exercise. Prove that, in Theorem 2.3, the assumption $\int_0^\infty xn^0(x)dx \geq \frac{r}{1-\alpha}$ is not needed because it will be true at some time t_0.

1. Assume $m_1(t) < \frac{r}{1-\alpha}$ for all times, then prove that all m_k are increasing, and that $m_1 \to \frac{r}{1-\alpha}$, as $t \to \infty$.

2. Using the equation on m_k prove they are bounded, and have a limit as $t \to \infty$.

3. Prove a contradiction with the formula for the solution $n(t, x)$ given in step (i).

2.4   Mutations

Until now we have seen ESS as the best adapted trait; in a population where initially all the traits are represented the ESS is the only one selected naturally.

We may ask what happens if a population initially does not represent all the traits but mutations may occur and create new mutants, that might be better adapted and thus be selected along with the selection principle we have already described, which themselves can mutate... etc The end of this chapter gives a formalism to address this question in rigorous mathematical terms.

In order to present this matter we need a more general framework than in Section 2.2, including a more general fitness in terms of the environment created by the total population. As we saw in previous sections, we can retain a structured population model that contains much of the previous examples while keeping some simplicity, writing

$$
\begin{cases}
\frac{d}{dt}n(t,x) = \big[b(x)Q_b(\varrho_b(t)) - d(x)Q_d(\varrho_d(t))\big]n(t,x), \quad x \in \mathbb{R},\, t \geq 0, \\[2mm]
\varrho_b(t) = \int_{\mathbb{R}} \psi_b(x)n(t,x)dx, \qquad \varrho_d(t) = \int_{\mathbb{R}} \psi_d(x)n(t,x)dx, \\[2mm]
n(t=0) = n^0(x) \geq 0.
\end{cases}
\tag{2.23}
$$

Here the notation is the same as before and we assume that $b, d, \psi_b, \psi_d \in C(\mathbb{R})$, $Q_b,\, Q_d \in C^1(\mathbb{R}^+)$ and there are real numbers such that

$$
Q_b'(\cdot) \leq 0, \qquad Q_d'(\cdot) \geq 0, \tag{2.24}
$$

$$
\begin{cases}
b_M \geq b(\cdot) \geq b_m > 0, \qquad d_M \geq d(\cdot) \geq d_m > 0, \\[2mm]
\psi_M \geq \psi_b(\cdot),\ \psi_d \geq \psi_m > 0,
\end{cases}
\tag{2.25}
$$

$$
\exists\, 0 < \varrho_m < \varrho_M \quad \text{such that} \quad
\begin{cases}
\max_{x \in \mathbb{R}} \big[b(x)Q_b(\psi_m \varrho_M) - d(x)Q_d(\psi_m \varrho_M)\big] < 0, \\[2mm]
\min_{x \in \mathbb{R}} \big[b(x)Q_b(\psi_M \varrho_m) - d(x)Q_d(\psi_M \varrho_m)\big] > 0.
\end{cases}
\tag{2.26}
$$

If reproduction is not completely faithful, an individual with trait y may generate offspring with trait x. Let $K(x,y)$ be the corresponding mutation kernel. To simplify, we only consider the case $K = K(x-y)$ with $K(\cdot)$ a probability density. Also the whole birth process should not necessarily give mutations but we postulate so in order to avoid additional parameters which would not change the heart of the matter.

To incorporate these mutations in (2.23), we postulate the dynamics given by

$$
\begin{cases}
\frac{\partial}{\partial t}n(t,x) = Q_b(\varrho_b(t)) \int_{\mathbb{R}} b(y)K(x-y)\, n(t,y)\, dy - d(x)Q_d(\varrho_d(t))n(t,x), \\[2mm]
\varrho_b(t) = \int_{\mathbb{R}} \psi_b(x)n(t,x)dx, \qquad \varrho_d(t) = \int_{\mathbb{R}} \psi_d(x)n(t,x)dx, \\[2mm]
n(t,0) = n^0(x) \geq 0.
\end{cases}
\tag{2.27}
$$

Our first claim is that this system is not much harder than an ordinary differential equation in terms of existence theory. We have

Theorem 2.4. *We assume (2.24)–(2.26), that $K(\cdot) \in L^1(\mathbb{R})$ is a probability density and $n^0 \in L^1$ satisfies $\varrho^m \leq \varrho^0 \leq \varrho^M$. Then the system (2.27) has a unique nonnegative solution such that $n, \frac{\partial}{\partial t} n \in C(\mathbb{R}^+; L^1(\mathbb{R}))$ and for all $t \geq 0$*

$$\varrho_m^0 \leq \varrho(t) \leq \varrho_M^0. \tag{2.28}$$

Proof. The method of proof is similar to that of the Cauchy–Lipschitz theorem and the main point is a global uniform bound. We begin with this bound and turn to the existence afterwards.

(i) A priori estimate. We first derive the inequalities (2.28) for nonnegative solutions. To do that, we integrate with respect to x the equation on n in (2.27) and notice that

$$\int_{\mathbb{R} \times \mathbb{R}} b(y)K(x-y)\, n(t,y)\, dy\, dx = \int_{\mathbb{R}} b(y)\, n(t,y)\, dy.$$

Therefore, we obtain

$$\frac{d}{dt}\varrho(t) \leq \int_{y \in \mathbb{R}} [b(y)Q_b(\psi_m \varrho(t)) - d(y)Q_d(\psi_m \varrho(t))]n(t,y)dy$$

$$\leq \varrho(t) \max_{y \in \mathbb{R}} \big[\, b(y)Q_b(\psi_m \varrho(t)) - d(y)Q_d(\psi_m \varrho(t)) \,\big].$$

The inequality from above in (2.28) follows directly since, whenever $\varrho(t)$ approaches ϱ_M (from below), then $\max_{y \in \mathbb{R}} \big[b(y)Q_b(\psi_m \varrho_M) - d(y)Q_d(\psi_m \varrho_M)\big]$ becomes negative and then $\varrho(t)$ decreases. A similar argument applies to prove the lower bound on $\varrho(t)$.

(ii) Existence. We prove existence using the Banach-Picard fixed point theorem. To do so, we consider the Banach space

$$X = C\big([0,T]; L^1(\mathbb{R})\big), \qquad \|m\|_X := \sup_{0 \leq t \leq T} \|m(t)\|_{L^1(\mathbb{R})},$$

for some $T > 0$ chosen later on. Then we define the closed subset

$$\mathcal{C} = \{m \in X,\ m \geq 0,\ \|m\|_X \leq C_\Phi\},$$

where C_Φ is defined, for T small enough, so that

$$\varrho_M + Tb_M Q_b(0)C_\Phi \leq C_\Phi,$$

and we can always impose for instance $C_\Phi \leq 2\varrho_M$.

For $m \in \mathcal{C}$, we define $R_b(t) = \int_{\mathbb{R}} \psi_b(x) m(t, x) dx$, $R_d(t) = \int_{\mathbb{R}} \psi_d(x) m(t, x) dx$ and the solution (x by x) to

$$
\begin{cases}
\frac{\partial}{\partial t} n(t, x) &= Q_b(R_b(t)) \int_{\mathbb{R}} b(y) K(x - y) \, m(t, y) \, dy - d(x) Q_d(R_d(t)) n(t, x), \\
n(t, 0) &= n^0(x) \geq 0.
\end{cases}
$$

(2.29)

Then, we claim that the operator

$$
m \mapsto \Phi(m) := n,
$$

has a unique fixed point in $\mathcal{C}$. We prove successively that

(a) $\Phi : \mathcal{C} \to \mathcal{C}$,

(b) Φ is a strong contraction for T small enough (depending only upon C_Φ).

Therefore we can apply the Banach–Picard theorem and deduce that Φ has a unique fixed point in $\mathcal{C}$ and the result is proved after iterating this argument on time-steps $[T, 2T], [2T, 3T], \ldots$. Notice also that, because this fixed point n belongs to X, $\frac{\partial}{\partial t} n$ also belongs to X, and in each time-step the solution satisfies, thanks to step (i), $\varrho_m \leq \varrho(t) \leq \varrho_M$.

The point (a) follows for example from the solution formula

$$
\begin{aligned}
n(t, x) =& n^0(x) e^{-d(x) \int_0^t Q_d(R_d)} \\
&+ Q_b(R_b) \int_{s=0}^t \int_{\mathbb{R}} b(y) K(x - y) \, m(s, y) \, dy \, e^{-d(x) \int_s^t Q_d(R_d)} ds,
\end{aligned}
$$

which implies the first two properties $n \geq 0$ and $n \in C([0, T]; L^1(\mathbb{R}))$. Then, we also have

$$
\frac{d}{dt} n(t, x) \leq Q_b(R_b(t)) \int_{\mathbb{R}} b(y) K(x - y) \, m(s, y) \, dy.
$$

Therefore we obtain for $t \leq T$,

$$
\begin{aligned}
\int_{\mathbb{R}} n(t, x) dx &\leq \int_{\mathbb{R}} n^0(x) dx + \int_0^t Q_b(R_b(s)) \int_{\mathbb{R}} b(y) \, m(s, y) \, dy \, ds \\
&\leq \varrho_M + T b_M Q_b(0) \sup_{0 \leq s \leq T} R(s) \\
&= \varrho_M + T b_M Q_b(0) C_\Phi \\
&= C_\Phi.
\end{aligned}
$$

This proves indeed that $\Phi(m)$ belongs to $\mathcal{C}$.

The point (b) is very standard. By substraction of solutions for two different m, say m^1 and m^2, we obtain

$$\frac{\partial}{\partial t}(n^1 - n^2) = [Q_b(R_b^1) - Q_b(R_b^2))]\int b(y)K(x-y)m^1(y)dy$$

$$+ Q_b(R_b^2)\int b(y)K(x-y)(m^1(y) - m^2(y))dy - d(x)[Q_d(R_d^1) - Q_d(R_d^2)]n^1$$

$$- d(x)Q_d(R_d^2)(n^1 - n^2).$$

From the a priori estimates of the step (a), the right-hand side is controlled by

$$C\|m^2 - m^1\|_{L^1(\mathbb{R})},$$

for some constant that uses C_Φ. Therefore, since the initial data for $n^2 - n^1$ vanishes,

$$\|m^2 - m^1\|_{L^1(\mathbb{R})} \le CT\|m^2 - m^1\|_{L^1(\mathbb{R})},$$

and we have obtained the strong contraction property for T small enough. Notice that as mentioned before, this time T is independent of n^0.   □

2.5   Small mutations: the constrained Hamilton–Jacobi equation

Along with the principles of adaptive dynamics, we now assume that mutations are small and we proceed to rescale accordingly the mutation model so as to arrive at a situation where a natural limit arises.

Hence, we introduce a (small) parameter $\varepsilon > 0$ in the mutation kernel and we set (and assume)

$$K_\varepsilon(z) = \frac{1}{\varepsilon}K(\frac{z}{\varepsilon}), \quad K(\cdot) \ge 0, \quad \int K = 1, \quad \int e^{z^2}K(z)dz < \infty. \qquad (2.30)$$

It is very clear that the dynamics described by the system (2.27) with K_ε in place of K is not interesting because the effect of mutations is too small. Indeed, we can write after the change of variable $y = x - \varepsilon z$,

$$\int_\mathbb{R} b(y)K_\varepsilon(x-y)\, n(t,y)\, dy = \int_\mathbb{R} b(x-\varepsilon z)K(z)\, n(t,x-\varepsilon z)\, dz$$

$$\xrightarrow[\varepsilon\to 0]{} \int_\mathbb{R} b(x)K(z)\, n(t,x)\, dz = b(x)\, n(t,x).$$

Therefore we end up with the original model (2.23).

To circumvent this, we observe the phenomena over a long time period, rescaling it with the same parameter ε. Hence we now consider the system

$$
\begin{cases}
\varepsilon \frac{\partial}{\partial t} n_\varepsilon(t,x) = Q_b(\varrho_{b,\varepsilon}(t)) \int_{\mathbb{R}} b(y) K_\varepsilon(x-y)\, n_\varepsilon(t,y)\, dy - Q_d(\varrho_{d,\varepsilon}(t)) n_\varepsilon(t,x), \\[2mm]
\varrho_{b,\varepsilon}(t) = \int_{\mathbb{R}} \psi_b(x) n_\varepsilon(t,x) dx, \qquad \varrho_{d,\varepsilon}(t) = \int_{\mathbb{R}} \psi_d(x) n_\varepsilon(t,x) dx, \\[2mm]
n_\varepsilon(t=0) = n_\varepsilon^0(x) \geq 0.
\end{cases}
$$

$$(2.31)$$

We explain below that, on this new time scale, a (or several in special cases) dominant trait $\bar{x}(t)$ is selected (following the selection principle of Sections 2.1 and 2.2). But now, this dominant trait evolves according to an interesting, and not explicit at all, dynamics. This means that, typically (but not always), we have

$$
n_\varepsilon(t,x) \xrightarrow[\varepsilon \to 0]{} \bar{\varrho}(t)\delta(x - \bar{x}(t)). \tag{2.32}
$$

Such a limiting population is called *monomorphic*. A polymorphic population corresponds to the sum of several Dirac masses in this limit. This arises for instance in a chemostat with several nutrients ([83]). We will refer to this (these) concentration point(s) as the dominant trait(s).

In order to derive the limiting dynamics, we assume a monomorphic initial population

$$
n_\varepsilon(t,x) = e^{\varphi_\varepsilon(t,x)/\varepsilon}, \qquad n_\varepsilon^0(x) = e^{\varphi_\varepsilon^0(x)/\varepsilon}, \tag{2.33}
$$

such that, as ε vanishes, there holds

$$
\begin{cases}
\int_{\mathbb{R}} n_\varepsilon^0(x) dx \to M^0 > 0, \qquad \varphi_\varepsilon^0 \to \varphi^0 \leq 0 \quad \text{uniformly in } \mathbb{R}, \\[2mm]
\max_{x \in \mathbb{R}} \varphi^0(x) = 0 = \varphi(\bar{x}^0) \quad \text{for a single } \bar{x}^0.
\end{cases} \tag{2.34}
$$

As an example, we might have in mind a gaussian distribution of traits in the population

$$
n_\varepsilon^0(x) = \frac{1}{\sqrt{2\pi\varepsilon}} e^{-|x|^2/2\varepsilon}, \qquad \varphi_\varepsilon^0(x) = -|x|^2 - \varepsilon \ln(2\pi\varepsilon).
$$

It clearly shows that the uniqueness of the maximum point of φ^0 is equivalent to the monomorphic Dirac concentration.

The reason why the change of unknown (2.33) is useful can be guessed from writing the equation on φ_ε:

$$
\begin{aligned}
\frac{\partial \varphi_\varepsilon(t,x)}{\partial t} &= e^{-\frac{\varphi_\varepsilon(t,x)}{\varepsilon}} Q_b(\varrho_{b,\varepsilon}(t)) \int_{\mathbb{R}} b(y) K_\varepsilon(x-y)\, e^{\frac{\varphi_\varepsilon(t,y)}{\varepsilon}}\, dy - Q_d(\varrho_{d,\varepsilon}(t)) d(x) \\[2mm]
&= Q_b(\varrho_{b,\varepsilon}(t)) \int_{\mathbb{R}} K(z)\, b(x+\varepsilon z)\, e^{\frac{\varphi_\varepsilon(t,x+\varepsilon z)-\varphi_\varepsilon(t,x)}{\varepsilon}}\, dz - Q_d(\varrho_{d,\varepsilon}(t)) d(x) \\[2mm]
&\xrightarrow[\varepsilon \to 0]{} Q_b(\varrho_b(t)) b(x) \int_{\mathbb{R}} K(z)\, e^{z \frac{\partial}{\partial x}\varphi(t,x)}\, dz - Q_d(\varrho_d(t)) d(x).
\end{aligned}
$$

Therefore the limiting function φ should satisfy the *constrained Hamilton–Jacobi equation*

$$
\begin{cases}
\frac{\partial}{\partial t}\varphi(t,x) = Q_b(\varrho_b(t))b(x)H\big(\frac{\partial}{\partial x}\varphi(t,x)\big) - Q_d(\varrho_d(t))d(x), & t \geq 0,\ x \in \mathbb{R}, \\
\max_{x\in\mathbb{R}} \varphi(t,x) = 0, & \forall t \geq 0, \\
\varphi(t=0,x) = \varphi^0(x) \leq 0
\end{cases}
$$

(2.35)

with

$$
H(p) := \int_{\mathbb{R}} K(z)\, e^{zp}\, dz.
$$

(2.36)

This function $H(\cdot)$ is called the *hamiltonian* and has the properties

$$
\begin{cases}
H(p) > 0, & H(0) = \int_{\mathbb{R}} K(z)\, dz = 1, & H''(p) = \int_{\mathbb{R}} K(z)\, z^2\, e^{zp}\, dz > 0, \\
H'(0) = \int_{\mathbb{R}} K(z)\, z\, dz = 0 & \text{for } K(\cdot) \text{ even,}
\end{cases}
$$

(2.37)

in other words, $H(\cdot)$ is a positive convex and even hamiltonian for K even. Finally the constraint $\max_{x\in\mathbb{R}} \varphi(t,x) = 0$ is equivalent to saying that n_e has bounded mass and non-extinction.

The problem (2.35)–(2.36) can be understood as follows. The unknowns are $\varphi(t,x)$ and the density $\bar\varrho(t)$ in (2.32). But they do not play the same role. The parameter $\bar\varrho(t)$ defines the *Lagrange multipliers* $Q_b(\bar\varrho_b(t))$ and $Q_d(\bar\varrho_d(t))$ for the constraint that $\max \varphi(t,\cdot) = 0$.

When the maximum of $\varphi(t,\cdot)$ is unique as initially (this is a monomorphic situation), we can recover the dominant trait in (2.32) by

$$
\max_{x\in\mathbb{R}} \varphi(t,x) = 0 = \varphi(t,\bar x(t)).
$$

Also the functions $Q_b(\bar\varrho_b(t))$ and $Q_d(\bar\varrho_d(t))$ are then given by

$$
\bar\varrho_b(t) = \psi_b(\bar x(t))\bar\varrho(t), \qquad \bar\varrho_d(t) = \psi_d(\bar x(t))\bar\varrho(t).
$$

More details are given at the end of Section 2.7 on the way to use these formulas.

The asymptotic, existence and uniqueness theory for deriving rigorously the constrained Hamilton-Jacobi equation (2.36) has been developed in [18] in the case $\psi_b = \psi_d$. In the general case uniqueness is an open question. The main difficulty is the interpretation of the two Lagrange multipliers $Q_b(\bar\varrho_b(t))$ and $Q_d(\bar\varrho_d(t))$ that makes it possible to contain the dimorphic situation of [83] where an additional uniqueness criteria is needed, and this is an open problem to find it. Again, we refer to Section 2.7 for more details.

We conclude this section by a remark: this asymptotic formalism is robust and can be extended to many situations under investigation. For instance it can be extended to systems as in [51] where a population with juveniles and adults is carried out, leading to a much more complicated hamiltonian.

2.6 A simple case of constrained Hamilton–Jacobi equation

A particularly simple situation is that of Section 2.1. Then $Q_b \equiv 1$, $d \equiv 1$ $\psi_d \equiv 1$ and $Q_d(u) = u$. Then the constrained Hamilton–Jacobi becomes

$$
\begin{cases}
\frac{\partial}{\partial t}\varphi(t,x) = b(x)H\left(\frac{\partial}{\partial x}\varphi(t,x)\right) - \bar{\varrho}(t), & t \geq 0,\ x \in \mathbb{R}, \\[2mm]
\max_{x \in \mathbb{R}} \varphi(t,x) = 0 = \varphi(t, \bar{x}(t)), & \forall t \geq 0, \\[2mm]
\varphi(t=0,x) = \varphi^0(x) \leq 0.
\end{cases}
\tag{2.38}
$$

Reflecting the simplicity of the model, this situation is particularly simple because one can reduce it to a more classical Hamilton–Jacobi equation. To do this, we introduce two new unknown functions

$$
R(t) = \int_0^t \varrho(s)ds, \qquad \psi(t,x) = \varphi(t,x) + R(t).
$$

Then, we arrive at the Hamilton–Jacobi equation (without a constraint) on ψ,

$$
\begin{cases}
\frac{\partial}{\partial t}\psi(t,x) = b(x)H\left(\frac{\partial}{\partial x}\psi(t,x)\right), & t \geq 0,\ x \in \mathbb{R}, \\[2mm]
\psi(t=0,x) = \varphi^0(x).
\end{cases}
\tag{2.39}
$$

This is the reason why one can derive a rigorous result that we state now but that we prove in Section 2.8. This is not possible for more complex dynamics as in [83], see also Section 2.9.

We use the notation

$$
R_\varepsilon(t) = \int_0^t \varrho_\varepsilon(s)ds, \qquad \psi_\varepsilon(t,x) = \varphi_\varepsilon(t,x) + R_\varepsilon(t).
\tag{2.40}
$$

Proposition 2.1. *Assume that* (2.30) *holds. We have, for all $t \geq 0$, the bounds*

$$
\min\left(\min_{y \in \mathbb{R}} b(y), \varrho_\varepsilon^0\right) \leq \varrho_\varepsilon(t) \leq \max\left(\max_{y \in \mathbb{R}} b(y), \varrho_\varepsilon^0\right).
\tag{2.41}
$$

If the initial data satisfies $\varphi_\varepsilon^0(x) \leq -|x| + C_\varepsilon^0$, then,

$$\psi_\varepsilon(t, x) \leq -|x| + C_\varepsilon^0 + t \ \max b \ \max_{|p| \leq 1} H(p), \tag{2.42}$$

$$\left|\frac{\partial}{\partial t}\psi_\varepsilon(t, x)\right| \leq \max_{y \in \mathbb{R}} |b(y)| \ \max_{y \in \mathbb{R}} H\left(\frac{\partial}{\partial y}\varphi_\varepsilon^0(y)\right). \tag{2.43}$$

Theorem 2.5. *We assume that (2.30), (2.34) hold and uniform bounds on the right hand sides in Proposition 2.1, on $\frac{\partial}{\partial x}\psi_\varepsilon^0(y)$ and on $\frac{\partial^2 b}{\partial x^2}$. Then, after extracting a subsequence, we have*

$$\psi_\varepsilon(t, x) \xrightarrow[\varepsilon \to 0]{} \psi(t, x), \quad \varphi_\varepsilon(t, x) \xrightarrow[\varepsilon \to 0]{} \varphi(t, x) \qquad \text{locally uniformly,}$$

ψ satisfies the Hamilton–Jacobi equation (2.39) in the sense of a viscosity solution and for all $t \geq 0$, $\max_{x \in \mathbb{R}} \varphi(t, x) = 0$. Also the inequalities in Proposition 2.1 hold for $\psi(t, x)$.

The notion of viscosity solutions for Hamilton–Jacobi equations has been introduced as a uniqueness criteria by Crandall and Lions (see [69] and the references therein). Its definition, motivation and main properties can be found in several recent surveys and books, see for instance [16, 69, 14, 98].

Theorem 2.5 gives a meaning to the system (2.35), simply the equation on φ has to be understood as an equation on ψ. Again, we refer to [18] for uniqueness.

2.7   A partial canonical equation and dynamics of the dominant trait

One can try to derive, from the Hamilton–Jacobi equation (2.35)–(2.36), an equation for the evolution of the dominant trait $\bar{x}(t)$. Such an equation, called *canonical equation* was proposed by Dieckmann and Law [72] for a model based on a stochastic process. We derive such an equation in our context and then apply it to monomorphic and dimorphic examples.

2.7.1   Derivation of a canonical equation

As we explain now, we can do it in our formalism but we can only derive partial information, namely the sense of variation of the trait. To do so, we assume here that

$$\frac{\partial H(0)}{\partial p} = 0, \qquad \text{i.e., } K(\cdot) \text{ is even.} \tag{2.44}$$

Our purpose is to prove

Lemma 2.1. *Assuming (2.44), in smoothness regimes, a dominant trait $\bar{x}(t)$ evolves according to the dynamics (canonical equation)*

$$Q_b(\bar{\varrho}_b(t))b(\bar{x}(t)) = Q_d(\bar{\varrho}_d(t))d(\bar{x}(t)), \tag{2.45}$$

$$\frac{d}{dt}\bar{x}(t) = \left(-\frac{\partial^2 \varphi(t,\bar{x}(t))}{\partial x^2}\right)^{-1}\left[Q_b(\bar{\varrho}_b(t))\frac{\partial b(\bar{x}(t))}{\partial x} - Q_d(\bar{\varrho}_d(t))\frac{\partial d(\bar{x}(t))}{\partial x}\right], \tag{2.46}$$

and because φ achieves a maximum at $\bar{x}(t)$, we have $\frac{\partial^2}{\partial x^2}\varphi(t,\bar{x}(t)) \le 0$.

In the terminology of adaptive dynamics, the quantity in the bracket is called the *selection gradient*.

Proof. Departing from the property that a dominant trait satisfies $\varphi(t,\bar{x}(t)) = 0$, and $\varphi(t,x) \le 0$, we also deduce that

$$\frac{\partial}{\partial t}\varphi(t,\bar{x}(t)) = 0, \qquad \frac{\partial}{\partial x}\varphi(t,\bar{x}(t)) = 0, \qquad \frac{\partial^2}{\partial x^2}\varphi(t,\bar{x}(t)) \le 0.$$

From the equation (2.35) we deduce directly (2.45).

We can also calculate

$$0 = \frac{d}{dt}\left[\frac{\partial}{\partial x}\varphi(t,\bar{x}(t))\right]$$

$$= \frac{\partial}{\partial t}\frac{\partial}{\partial x}\varphi(t,\bar{x}(t)) + \dot{\bar{x}}(t)\frac{\partial^2}{\partial x^2}\varphi(t,\bar{x}(t)).$$

Next, we differentiate in x the equation (2.35) and arrive at

$$\frac{\partial^2 \varphi(t,x)}{\partial t\, \partial x} = Q_b(\bar{\varrho}_b(t))\frac{\partial b(x)}{\partial x}H\left(\frac{\partial}{\partial x}\varphi\right) + b(x)H_p\left(\frac{\partial}{\partial x}\varphi\right)\frac{\partial^2 \varphi(t,x)}{\partial x^2} - Q_d(\bar{\varrho}_d(t))\frac{\partial d(x)}{\partial x}.$$

At the point $\bar{x}(t)$ we have, along with (2.44), $\frac{\partial H(0)}{\partial p} = 0$ and thus we obtain simply

$$\frac{\partial}{\partial t}\frac{\partial}{\partial x}\varphi(t,\bar{x}(t)) = Q_b(\bar{\varrho}_b(t))\frac{\partial b(\bar{x}(t))}{\partial x} - Q_d(\bar{\varrho}_d(t))\frac{\partial d(\bar{x}(t))}{\partial x}.$$

With the above information on $\dot{\bar{x}}(t)$ we arrive at (2.46). □

Notice that one cannot close this by another equation on $\frac{\partial^2 \varphi(t,\bar{x}(t))}{\partial x^2}$. Indeed such an equation involves a higher order derivative $\frac{\partial^3 \varphi(t,\bar{x}(t))}{\partial x^3}$ and so on. Also, it might turn out that $\frac{\partial^2 \varphi(t,\bar{x}(t))}{\partial x^2}$ vanishes or that φ achieves several maxima at some times. These are singularities where the canonical equation does not apply.

Nevertheless we can infer from the equation (2.46) the dynamics of the dominant trait and we distinguish the monomorphic and dimorphic cases.

2.7.2　Monomorphic adaptive evolution

We assume that the population is monomorphic, i.e., there is a single trait $\bar{x}(t)$, at each time, such that $\varphi(t, \bar{x}(t)) = 0$, or in other words that the limit of n_ε is a single Dirac mass. Then, we obtain in (2.31) that

$$\bar{\varrho}_b(t) = \bar{\varrho}(t)\psi_b(\bar{x}(t)), \qquad \bar{\varrho}_d(t) = \bar{\varrho}(t)\psi_d(\bar{x}(t)).$$

Therefore, using (2.45)

$$Q_b\big(\bar{\varrho}(t)\psi_b(\bar{x}(t))\big)b(\bar{x}(t)) = Q_d\big(\bar{\varrho}(t)\psi_d(\bar{x}(t))\big)d(\bar{x}(t)) \quad \forall t \geq 0.$$

This relation defines (say by continuity) $\bar{\varrho}(t)$ as a function $R(\bar{x}(t))$ which can be inserted in (2.46) to obtain a differential equation on $\bar{x}(t)$ (once $D^2\varphi$ is known).

At least, we can use the canonical equation (2.46) to deduce that $\bar{x}(t)$ evolves in the direction of increasing values in x of the quantity

$$Q_b(\bar{\varrho}_b(t))b(x) - Q_d(\bar{\varrho}_d(t))d(x).$$

This explains the denomination *selection gradient* for its x derivative. Therefore, $\big(\bar{\varrho}_b(t), \bar{\varrho}_d(t), \bar{x}(t)\big)$ will generically reach a point $\big(\bar{\varrho}_b^*, \bar{\varrho}_d^*, \bar{x}^*\big)$, where

$$0 = Q_b(\bar{\varrho}_b^*)b(\bar{x}^*) - Q_d(\bar{\varrho}_d^*)d(\bar{x}^*) = \max_{x \in \mathbb{R}} \big[Q_b(\bar{\varrho}_b^*)b(x) - Q_d(\bar{\varrho}_d^*)d(x)\big], \qquad (2.47)$$

with

$$\bar{\varrho}_b^* = \psi_b(\bar{x}^*)\bar{\varrho}^*, \qquad \bar{\varrho}_d^* = \psi_d(\bar{x}^*)\bar{\varrho}^*. \qquad (2.48)$$

These two formulas (2.47)–(2.48) on the two unknowns $\bar{x}^*$ and $\bar{\varrho}^*$ are typical characterizations of the ESS as we have already seen it rigorously in specific examples (see Sections 2.1 and 2.2).

We recall that the logic behind these formula has been explained in Section 2.2, when ψ_b and ψ_d are constant. Then the function

$$\rho \mapsto \max_{x \in \mathbb{R}} \big[Q_b(\psi_b\bar{\varrho})b(x) - Q_d(\psi_d\bar{\varrho})d(x)\big]$$

is monotonic (when Q_b is decreasing and Q_d is increasing) and can vanish at a single point $\bar{\varrho}^*$. Then, if b/d is monotonic, then the $\max_x$ is attained itself at a single point x^*.

2.7.3　Dimorphic adaptive evolution

A dimorphic population can occur if there are two dominant traits

$$n_\varepsilon(t, x) \to \bar{\varrho}_1(t)\delta(x - \bar{x}_1(t)) + \bar{\varrho}_2(t)\delta(x - \bar{x}_2(t)),$$

which also means

$$0 = \max_{x} \varphi(t, x) = \varphi(t, \bar{x}_1(t)) = \varphi(t, \bar{x}_2(t)).$$

According to the relation (2.45) this means that we should have

$$\frac{Q_b(\bar{\varrho}_b(t))}{Q_d(\bar{\varrho}_d(t))} = \frac{d(\bar{x}_1(t))}{b(\bar{x}_1(t))} = \frac{d(\bar{x}_2(t))}{b(\bar{x}_2(t))} := R(t).$$

In other words, a condition for dimorphism is that b/d is not one-to-one. If it has a 'parabolic profile', the above ratio $R(t)$ determines $\bar{x}_1(t)$ and $\bar{x}_2(t)$. Which themselves constrain $\varrho_1(t)$ and $\varrho_2(t)$ by a first relation

$$Q_b\big(\bar{\varrho}_1(t))\psi_b(\bar{x}_1(t)) + \bar{\varrho}_2(t))\psi_b(\bar{x}_2(t))\big) = R(t)Q_b\big(\bar{\varrho}_1(t))\psi_b(\bar{x}_1(t)) + \bar{\varrho}_2(t))\psi_b(\bar{x}_2(t))\big).$$

This also means that the two differential equations given by (2.46) are in fact reduced to the same dynamics (say for instance on $\bar{x}_1(t)$) through the inverse of the second derivative of φ.

The evolution of two such traits is therefore highly constrained and it is not as easy as before to characterize the dimorphic ESS. The same logic leads to

$$0 = \max_{x} \big[Q_b(\bar{\varrho}_b^*)b(x) - Q_d(\bar{\varrho}_d^*d(x)\big] = Q_b(\bar{\varrho}_b^*)b(\bar{x}_i^*) - Q_d(\bar{\varrho}_d^*)d(x_i^*), \qquad i = 1, \, 2,$$

$$\bar{\varrho}_b^* = \bar{\varrho}_1^*\psi_b(x_1^*) + \bar{\varrho}_2^*\psi_b(x_2^*),$$

$$\bar{\varrho}_d^* = \bar{\varrho}_1^*\psi_d(x_1^*) + \bar{\varrho}_2^*\psi_d(x_2^*).$$

Again, we can analyze these formulae in the case where ψ_b and ψ_d are constant (and these constants should be different!). Then $\bar{\varrho}_b^*$ and $\bar{\varrho}_d^*$ are two independent Lagrange multipliers and we may expect a one-parameter family of dimorphic ESS.

Also, the constrained Hamilton–Jacobi equation (2.35) is clearly not enough to decide between the monomorphic interpretation and the dimorphic equation and this is a reason for non-uniqueness. In the monomorphic case $\varphi(t, \cdot)$ achieves a single maximum with a single Lagrange multiplier in the interpretation $n(t, x) = \varrho(t)\delta(x - \bar{x}(t))$. In the dimorphic case $\varphi(t, \cdot)$ achieves two maxima that require two independent Lagrange multipliers provided by the interpretation $n(t, x) = \varrho_1(t)\delta(x - \bar{x}_1(t)) + \varrho_2(t)\delta(x - \bar{x}_2(t))$. It is an open problem to find a criterion for uniqueness.

2.7.4 Adaptive evolution: cannibalism

As an example we can treat the simple model of cannibalism from Section 2.3. This corresponds, after rescaling and introducing mutations, to the equation

$$
\begin{cases}
\varepsilon \frac{\partial}{\partial t} n_\varepsilon(t,x) = \int_{\mathbb{R}} \big(r + \alpha y \varrho_\varepsilon(t)\big) K_\varepsilon(x-y)\, n_\varepsilon(t,y)\, dy - \varrho_{d,\varepsilon}(t) n_\varepsilon(t,x), \\[2mm]
\varrho_\varepsilon(t) = \int_{\mathbb{R}} n_\varepsilon(t,x) dx, \qquad \varrho_{d,\varepsilon}(t) = \int_{\mathbb{R}} x n_\varepsilon(t,x) dx, \\[2mm]
n_\varepsilon(t=0) = n_\varepsilon^0(x) \geq 0.
\end{cases}
\tag{2.49}
$$

The limit $\varepsilon \to 0$ gives the constrained Hamilton–Jacobi equation

$$
\frac{\partial}{\partial t}\varphi(t,x) = \big(r + \alpha x \varrho(t)\big) H(\nabla \varphi) - \varrho_d(t).
\tag{2.50}
$$

Therefore the dominant trait should satisfy the relation (2.45) which here reads

$$
r + \alpha \bar{x}(t)\varrho(t) = \varrho_d(t).
$$

Since there is a single possible solution $\bar{x}(t)$, the population can only be monomorphic (see [79] for more elaborate models with possibly dimorphism). Also as we already had it intuitively from Section 2.3, the canonical equation (2.46) gives the evolution

$$
\frac{d}{dt}\bar{x}(t) = \Big(-\frac{\partial^2 \varphi(t,\bar{x}(t))}{\partial x^2}\Big)^{-1} \alpha \varrho(t) > 0,
$$

(here the *selection gradient* is simply $\alpha \varrho(t)$) which tells us that the trait increases (to infinity in principle).

2.7.5 Adaptive evolution: chemostat with two nutrients

In the case of two nutrients with simplified predation function and for fast response of the nutrients, the system (2.1) becomes

$$
\begin{cases}
S_1(t) = \frac{S_{01}}{1+\int \eta_1(x) n(t,x) dx}, \qquad t \geq 0,\ x > 0, \\[2mm]
S_2(t) = \frac{S_{02}}{1+\int \eta_2(x) n(t,x) dx}, \\[2mm]
\frac{\partial}{\partial t} n(t,x) = -R n(t,x) + \int \big(\eta_1(y) S_1(t) + \eta_2(y) S_2(t)\big) n(t,y) K_\varepsilon(x-y) dy, \\[2mm]
n(t=0,x) = n^0(x) > 0,\ n^0 \in L^1 \cap L^\infty(\mathbb{R}^+).
\end{cases}
\tag{2.51}
$$

One can deduce the constrained Hamilton–Jacobi equation

$$
\frac{\partial}{\partial t}\varphi(t,x) = -R + \big(\eta_1(x) S_1(t) + \eta_2(x) S_2(t)\big) H(\nabla \varphi).
$$

In the present case the *selection gradient* is $\eta_1'(x)S_1(t) + \eta_2'(x)S_2(t)$.

Therefore the dominant traits should satisfy the relation (2.45) which here reads

$$R = \eta_1(x_1(t))S_1(t) + \eta_2(x_1(t))S_2(t) = \eta_1(x_2(t))S_1(t) + \eta_2(x_2(t))S_2(t).$$

It is possible to achieve dimorphism if the function $\eta_1(x)S_1 + \eta_2(x)S_2$ has a 'parabolic profile'. This is possible for instance when η_j are positive convex functions with η_1 decreasing and η_2 increasing.

2.8 A rigorous derivation of the Hamilton–Jacobi equation

This section is devoted to the proofs of the results stated in Section 2.5.

Proof of Proposition 2.1. The first statement, (2.41), is nothing but the a priori estimate (2.28). For the second statement, (2.42), we write the Hamilton–Jacobi equation on ψ_ε

$$\begin{cases} \frac{\partial}{\partial t}\psi_\varepsilon(t,x) & = \int b(x+\varepsilon z)e^{(\psi_\varepsilon(t,x+\varepsilon z)-\psi_\varepsilon(t,x))/\varepsilon}M(z)dz := K_\varepsilon\big(x,\psi_\varepsilon(t,\cdot)\big), \\ \psi_\varepsilon(t=0,x) & = \varphi_\varepsilon^0(x). \end{cases}$$

$$(2.52)$$

We just notice that $\bar\psi_\varepsilon(t) := -|x| + C_\varepsilon^0 + t \,\max b \,\max_{|p|\leq 1} H(p)$ is a supersolution to this Hamilton–Jacobi equation (and also to (2.39)) since

$$\frac{\partial}{\partial t}\bar\psi_\varepsilon(t,x) = \max b \,\max_{|p|\leq 1} H(p) \geq K_\varepsilon\big(x,\bar\psi(t,\cdot)\big) \qquad \text{and} \;\geq b(x)H\big(\frac{\partial}{\partial x}\bar\psi(t,x)\big).$$

For the last statement, (2.43), we consider $\Psi_\varepsilon(t,x) = \frac{\partial}{\partial t}\psi_\varepsilon(t,x)$ and differentiate the equation (2.52) to obtain

$$\frac{\partial}{\partial t}\Psi_\varepsilon(t,x) = \frac{1}{\varepsilon}\int b(x+\varepsilon z)e^{(\psi_\varepsilon(t,x+\varepsilon z)-\psi_\varepsilon(t,x))/\varepsilon}\big(\Psi_\varepsilon(t,x+\varepsilon z) - \Psi_\varepsilon(t,x)\big)M(z)dz.$$

At the maximum point, $\Psi_\varepsilon(t,x_0) = \max_{x\in\mathbb{R}}\Psi_\varepsilon(t,x)$, we deduce that

$$\frac{\partial}{\partial t}\Psi_\varepsilon(t,x_0) \leq 0,$$

and thus $\max_{x\in\mathbb{R}}\Psi_\varepsilon(t,x) \leq \max_{x\in\mathbb{R}}\Psi_\varepsilon(t=0,x)$. The same argument leads to control the *minimum*. Finally, we use the equation (2.52) at $t=0$ to obtain

$$|\Psi_\varepsilon(t=0,x)| = \int b(x+\varepsilon z)e^{(\varphi_\varepsilon^0(x+\varepsilon z)-\varphi_\varepsilon^0(x))/\varepsilon}M(z)dz$$

$$\leq \max_{y\in\mathbb{R}}|b(y)| \,\max_{y\in\mathbb{R}} H\big(\frac{\partial}{\partial x}\varphi_\varepsilon^0(y)\big),$$

and the third statement is proved. □

Proof of Theorem 2.5. As a first step, we now derive a bound on the x derivative of ψ_ε. To do so we fix a time $T > 0$ and set $\Phi_\varepsilon(t, x) = \frac{\partial}{\partial x}\psi_\varepsilon(t, x)$. We write the equation (2.52) as

$$\frac{\partial}{\partial t}\frac{\psi_\varepsilon(t, x)}{b(x)} = \int \frac{b(x + \varepsilon z)}{b(x)} e^{(\psi_\varepsilon(t, x + \varepsilon z) - \psi_\varepsilon(t, x))/\varepsilon} M(z)dz.$$

Next, we compute

$$\begin{aligned}
\frac{\partial}{\partial t}\frac{\Phi_\varepsilon(t, x)}{b(x)} = & \frac{\partial}{\partial t}\psi_\varepsilon(t, x)\frac{\partial b(x)}{\partial x}/b(x)^2 \\
& + \int \frac{\partial}{\partial x}\left(\frac{b(x + \varepsilon z)}{b(x)}\right) e^{(\psi_\varepsilon(t, x + \varepsilon z) - \psi_\varepsilon(t, x))/\varepsilon} M(z)dz \\
& + \frac{1}{\varepsilon}\int \frac{b(x + \varepsilon z)}{b(x)} e^{(\psi_\varepsilon(t, x + \varepsilon z) - \psi_\varepsilon(t, x))/\varepsilon}\left(\Phi_\varepsilon(t, x + \varepsilon z) - \Phi_\varepsilon(t, x)\right) M(z)dz.
\end{aligned}$$

Again, we consider the maximum point

$$\Phi_\varepsilon(t, x_0) = \max_{y \in \mathbb{R}}|\Phi_\varepsilon(t, y)| := Q_\varepsilon(t),$$

and estimate

$$\left|\frac{\partial}{\partial x}\left(\frac{b(x + \varepsilon z)}{b(x)}\right)\right| \le \varepsilon\left[\max_{y \in \mathbb{R}}\left|\frac{\partial^2 b(y)}{\partial y^2}\right| \max_{y \in \mathbb{R}}|b(y)| + \max_{y \in \mathbb{R}}\left|\frac{\partial b(y)}{\partial y}\right|^2\right]/b(x)^2.$$

Multiplying by $b(x)$ we deduce again, because at the maximum point the last term in the right hand-side is non-positive, and at the minimum point it is nonnegative, that

$$\frac{d}{dt}Q_\varepsilon(t) \le \max_{y \in \mathbb{R}}\left[\left|\frac{\partial}{\partial t}\varphi_\varepsilon^0(y)\right|\left|\frac{\partial b(y)}{\partial x}/b(y)\right|\right] + \varepsilon\, Q_b\, e^{Q_\varepsilon(t)}\int M(z)dz$$

with $Q_b = [\max_{y \in \mathbb{R}}|\frac{\partial^2 b(y)}{\partial y^2}|\ \max_{y \in \mathbb{R}}|b(y)| + \max_{y \in \mathbb{R}}|\frac{\partial b(y)}{\partial y}|^2]/[\min_{y \in \mathbb{R}} b(y)]$. This provides a uniform bound on $Q_\varepsilon(t)$ which proves that ψ_ε is locally compact thanks to the Ascoli–Arzela Theorem.

Therefore we may extract subsequences such that $R_\varepsilon(t)$ (which is uniformly lipschitz continuous) and $\psi_\varepsilon(t, x)$ converge locally uniformly. This is enough to pass to the limit in the viscosity sense in the equation (2.52) and to recover (2.39) (see [16, 69, 14, 98] for basic handling of the Hamilton–Jacobi equation). Finally, the statement $\max_{x \in \mathbb{R}}\varphi(t, x) = 0$ follows from the bound (2.41) which would not hold if $\max_{x \in \mathbb{R}}\varphi(t, x)$ would be positive or negative in view of the definition $n_\varepsilon = e^{\varphi_\varepsilon/\varepsilon}$. $\qquad\qquad\square$

2.9 Numerical results and examples of adaptive evolution

In this section we present several numerical experiments that illustrate possible evolutions of the dominant traits and that allow comparison between the direct simulation of the structured population equations (2.31) and the constrained Hamilton–Jacobi equation (2.35). The advantage of the latter is that it requires very few points compared to the former which require several mesh points per ε.

We first present several possible behaviors for monomorphic situations. They correspond to the general formalism (2.35) with $Q_d \equiv 1$ and $Q_b = \varrho = \int n(t, x)dx$ with a trait which is chosen, for numerical treatment in $[0, 1]$, and we present various cases of function $b(x)$. A continuous evolution is presented in Figure 2.1 for the function $b(x) = .5 + x(2 - x)$.

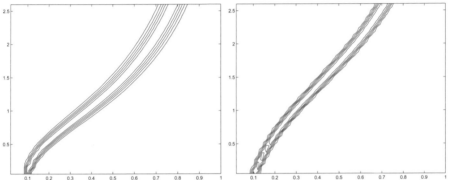

Figure 2.1: Continuous evolution of a trait according to direct simulation of the structured population equations (2.31) (left) and the constrained Hamilton–Jacobi equation (2.35) (right). The horizontal coordinate is the trait x and the vertical coordinate is time.

In Figure 2.2 we illustrate the possibility of a discontinuity of the dominant trait and this corresponds to the choice $b(x) = \min(.45 + x^2, .55 + .4x)$, with the same values of Q_b and Q_d as before. The main interest of this example is to show that a mere differential equation cannot be enough to describe the evolution of the dominant trait.

Finally in Figure 2.3, we show the bifurcation from a monomorphic to a dimorphic population in the case of the chemostat with two nutrients (2.51), following the theory and example developed in [83].

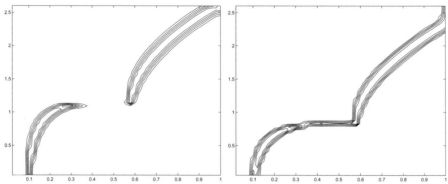

Figure 2.2: SAME AS FIGURE 2.1 FOR A TRAIT EVOLUTION WITH A JUMP IN A MONOMORPHIC POPULATION (THIS SHOWS THAT A DIFFERENTIAL EQUATION CANNOT BE ENOUGH TO DESCRIBE THE EVOLUTION OF THE DOMINANT TRAIT).

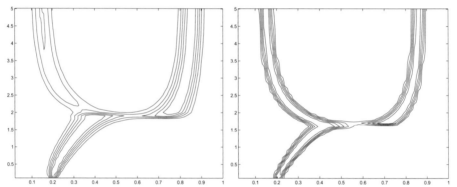

Figure 2.3: SAME AS FIGURE 2.1 FOR A TRAIT EVOLUTION WHICH EXHIBITS BRANCHING. THIS IS THE CASE OF A CHEMOSTAT WITH TWO NUTRIENTS.

Another and in some sense opposite example of possible evolution for a dimorphic population, is that it evolves toward a monomorphic situation because the ESS is attractive for both monomorphic and dimorphic populations. We borrow the Figure 2.4 from [51] where a system for adults and juveniles is treated.

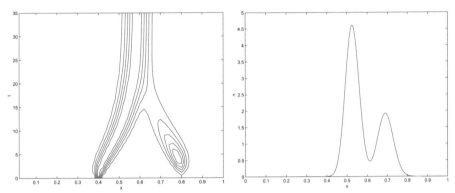

Figure 2.4: A DIMORPHIC POPULATION BECOMES MONOMORPHIC. TAKEN FROM [51] WHO CONSIDER A SYSTEM FOR ADULTS AND JUVENILES.

Chapter 3

Population balance equations: the renewal equation

In the next chapters we focus our attention on particular models of physiologically structured population dynamics where the structure leads to a Partial Differential Equation (compared to the models in Sections 2.4 and 1.5.3 for instance). Therefore they involve a different mathematical formalism. But these models also differ by their biological meaning; the variable x is no longer a trait inherited from the birth of individuals, but a variable (age structure, size structure, maturity or maturation structure) that can evolve all along their life.

Another typical property is that these models that are written so as to combine in a non-conservative way, various conservation laws (total mass, total number of individuals). Also additional terms (death, maturation speed, mixing of several conservation laws) makes the overall balance law not visible (compared to the case of the chemostat in Section 1.2.3 for instance). We refer to such models as *population balance laws* by analogy with the conservation laws of physics or continuum mechanics (see [71]). A natural question is: how should the mathematical tools for these models be modified to take into account these birth and death processes?

An outcome of our presentation is that a common concept with those models arising in physics, namely *entropy*, plays a role here and can be extended to what we call *generalized relative entropy*. This kind of model describes unrestricted growth of the population and thus a Malthus parameter gives the exponential growth. Once renormalized by this exponential factor, the entropy dissipation drives the system to a steady state that minimizes the entropy. Of course this conclusion is at odds with observations. Biological systems exhibit a strong variability (certainly due to inherent mutations and adaptation facilities as modeled by adaptive dynamics in Chapter 2). But the explanation is simple. Here, the regime is described by linear equations and thus as mentioned earlier it is limited to the initial unrestricted growth, when a species is introduced in a favorable environment without nutrients limitation, neither competition nor selection. We refer to Section 6.4 for the general concept of Generalized Relative Entropy (following [176]) in the framework of second order PDEs or integral equations.

As a warm-up, this chapter deals with a first example, we consider that the population under consideration is only physiologically structured by age, following the so-called McKendrick-Von Foerster equation or renewal equation (see [9, 162, 230, 171] for instance). The model was originally introduced for epidemiology and is now commonly used in demographic studies, however with more ingredients than we use here (immigration, sex, cultural classes, etc.) but it is also used in various other applications, [64]. The next chapter deals with the more interesting case of size structure.

3.1   The renewal equation

Consider a population density $n(t, x) \geq 0$ with an age structure $x \in (0, \infty)$ at time $t \in (0, \infty)$ which is only subject to aging and birth, ignoring the death rate for the time being (see the exercise for an extension with a death rate). Denoting by $B(x)$ the birth rate, we arrive at the equation

$$\begin{cases} \frac{\partial}{\partial t}n(t,x) + \frac{\partial}{\partial x}n(t,x) = 0, \quad t \geq 0, \, x \geq 0, \\[2mm] n(t, x = 0) = \int B(y)n(t,y)dy, \\[2mm] n(t = 0, x) = n^0(x). \end{cases} \tag{3.1}$$

We give here a method to analyze the existence, the main properties and the long time behavior of (3.1) based on the general relative entropy method. More precisely, we use the method developed in [177]. We also point out that the equation (3.1) is a special case of the cell division equation (4.16), but it is the only case where all calculations can be performed explicitly and thus allows very easy analysis. This explains that it has been commonly used either using the method of characteristics or following a Laplace transform method due to Feller [100].

The method we develop here is easier, more general because we do not require that $B(x) = 0$ for x larger than some $x^\sharp$ and expresses results in precise weighted norms. To illustrate it, we just use the assumptions

$$B(\cdot) \geq 0, \qquad B \in L^\infty(\mathbb{R}^+), \qquad 1 < \int_0^\infty B(x)dx < +\infty. \tag{3.2}$$

See Sections 3.8, 3.9 for extensions of this assumption.

3.2   Eigenelements

Before we state our existence result for the age structured model (3.1), we need some notation and equations that are used later to state natural results. First of all, we look for (λ_0, N, ϕ), first eigenelements associated to the renewal equation (3.1). These are defined (see Section 6.4) by

$$
\left\{
\begin{array}{ll}
\frac{\partial}{\partial x} N(x) + \lambda_0 \, N(x) = 0, & x \geq 0, \\[2mm]
N(0) = \int B(y) N(y) dy, & \\[2mm]
N(\cdot) > 0, \quad \int N(y) dy = 1, &
\end{array}
\right. \tag{3.3}
$$

$$
\left\{
\begin{array}{ll}
-\frac{\partial}{\partial x} \phi(x) + \lambda_0 \, \phi(x) = \phi(0) B(x), & x \geq 0, \\[2mm]
\phi(\cdot) \geq 0, \quad \int N(y) \phi(y) dy = 1. &
\end{array}
\right. \tag{3.4}
$$

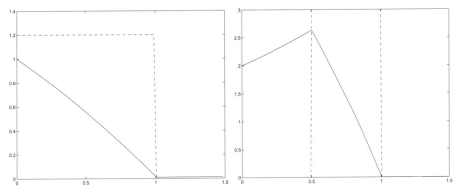

Figure 3.1: SOLUTION ϕ OF THE DUAL EIGENVALUE PROBLEM, NORMALIZED DIFFERENTLY THAN IN (3.4), CONTINUOUS LINE. THE DASHED LINE IS B. LEFT: $B(x) = 1.2 \, \mathbf{1}_{\{x<1\}}$. RIGHT: $B(x) = 3 \, \mathbf{1}_{\{.5 \leq x \leq 1\}}$.

The function N is the eigenvector associated with the operator in (3.1) and ϕ the eigenvector associated with the adjoint operator. Notice that there is no boundary condition for equation (3.4), because it should be given at $x = \infty$ and it is replaced by the positivity and integrability conditions. The parameter λ_0 is also called the *Malthus parameter*.

The existence is easy thanks to explicit solutions as we state it now.

Lemma 3.1. *Under assumptions* (3.2) *there is a unique solution* $(\lambda_0 > 0, N, \phi)$ *to equations* (3.3)–(3.4) *and* $\phi(x) \leq \|B\|_{L^\infty} / [\lambda_0 \int y B(y) e^{-\lambda_0 y} dy]$.

Proof. First step. The unique solution to the equation for N gives the formula

$$
N(x) = \lambda_0 e^{-\lambda_0 x}. \tag{3.5}
$$

We have to check that it satisfies the boundary condition, i.e.,

$$
\int B(x) e^{-\lambda_0 x} dx = 1. \tag{3.6}
$$

This is possible, with the assumptions (3.2), because the function $\lambda \mapsto B(x)e^{-\lambda x}$ is integrable for all $\lambda > 0$ and thus continuous, it is clearly deceasing, and by the Lebesgue monotone convergence theorem, we have

$$\lim_{\lambda \to 0^+} \int B(x)e^{-\lambda x}dx > 1, \qquad \lim_{\lambda \to +\infty} \int B(x)e^{-\lambda x}dx = 0.$$

Therefore there is a unique value λ_0 which fulfills the condition (3.6).

Second step. The equation on ϕ can also be written on $Q(x) = \phi(x)N(x)/[\phi(0)N(0)]$ as

$$\begin{cases} \frac{\partial}{\partial x}Q(x) = -B(x)N(x)/N(0), & x \geq 0, \\ \\ Q(x=0) = 1, & Q \geq 0, \quad \int Q(x)dx < \infty, \end{cases} \qquad (3.7)$$

and the choice of $\phi(0)$ then allows the normalization $\int N\phi = 1$. It has a unique solution, thanks to (3.5) and the condition (3.6), given by

$$Q(x) = 1 - \int_0^x B(y)e^{-\lambda_0 y}dy = \int_x^\infty B(y)e^{-\lambda_0 y}dy.$$

And we notice that

$$0 \leq Q(x) \leq 1.$$

Third step. The normalisation $\int N\phi = 1$ is also possible and gives (because B is bounded) the choice of $\phi(0)$. We write, using the Fubini theorem,

$$\frac{1}{N(0)\phi(0)} = \int Q(x)dx = \int_0^\infty yB(y)e^{-\lambda_0 y}dy.$$

Therefore

$$\frac{\phi(x)N(x)}{N(0)\phi(0)} = Q(x) \leq \|B\|_{L^\infty} \int_x^\infty e^{-\lambda_0 y}dy = \frac{\|B\|_{L^\infty}}{\lambda_0}e^{-\lambda_0 x},$$

and the upper bound on ϕ follows from $N(0) = \lambda_0$. $\qquad\qquad\qquad\qquad\qquad \square$

Notice that the solution N is trivial, but ϕ can have a more complicated shape. Two examples are depicted in Figure 3.1. Especially, when B has a compact support, ϕ has also a compact support (the convex hull of the support of B and $\{0\}$).

Notice also that the threshold 1 in the integral condition of (3.2) is sharp in order to ensure an expanding population (but the theory does not use it so much). See the calculation leading to formula (3.6) of λ_0.

3.3   Existence theory

We are now ready for our first existence result where it is convenient to use the notation
$$\widetilde{n}(t,x) = n(t,x)e^{-\lambda_0 t}.$$
The function $\widetilde{n}$ satisfies

$$
\begin{cases}
\frac{\partial}{\partial t}\widetilde{n}(t,x) + \frac{\partial}{\partial x}\widetilde{n}(t,x) + \lambda_0\,\widetilde{n}(t,x) = 0, & t \geq 0,\ x \geq 0, \\[2mm]
\widetilde{n}(t,x=0) = \int B(y)\widetilde{n}(t,y)dy, \\[2mm]
\widetilde{n}(t=0,x) = n^0(x).
\end{cases}
\tag{3.8}
$$

Theorem 3.1. *Under assumptions (3.2) and for an initial data satisfying*
$$\exists C_0, \qquad |n^0(x)| \leq C_0 N(x),$$
there is a unique solution in distribution sense $\widetilde{n} \in C\big(\mathbb{R}^+; L^1(\mathbb{R}^+; \phi(x)dx)\big)$ *to (3.8) and we have*

(i) *the maximum principle*
$$|\widetilde{n}(t,x)| \leq C_0 N(x) \qquad \forall t \geq 0,$$

(ii) *the comparison principle*
$$n_1^0 \leq n_2^0 \quad \Rightarrow \quad \widetilde{n}_1(t,x) \leq \widetilde{n}_2(t,x),$$

(iii) *the conservation law and the* $L^1(\phi(x)dx)$ *contraction principle hold,*
$$\int_0^\infty \widetilde{n}(t,x)\phi(x)dx = \int_0^\infty n^0(x)\phi(x)dx,$$
$$\int_0^\infty |\widetilde{n}(t,x)|\phi(x)dx \leq \int_0^\infty |n^0(x)|\phi(x)dx.$$

Remark 3.1. The interested reader can check a curiosity here which shows that the renewal equation is not completely evident. Under the assumption (3.2) and
$$\int |n^0(x)|\,\phi(x)dx < \infty,$$
there is a weak solution $\widetilde{n} \in C\big(\mathbb{R}^+; L^1(\mathbb{R}^+; \phi(x)dx)\big)$ to the equation written on $\widetilde{n}\phi$. This solution is built as the limit of the Cauchy sequence as in the proof below. Surprisingly, it can satisfy $\widetilde{n}(t,0) = \infty$, i.e., $\int_\mathbb{R} B(x)\widetilde{n}(t,x)dx = +\infty$ at $t = 0$ but a regularizing effect (in fact closer to a hypercontractivity effect) occurs. We have
$$\int_0^T \int_0^\infty B(x)|\widetilde{n}(t,x)|[\psi(0) - \psi(x)]dx\,dt \leq C(T,\psi) < \infty,$$

for all decreasing, lipschitz continuous test functions ψ. To prove it, choose simply the test function ψ in the equation on $|\tilde{n}|\ \phi$ (see below) and notice that, because $\int |n(t,x)|\ \phi(x)dx \leq \int |n^0(x)|\ \phi(x)dx$, all the other terms are under control because they contain $\phi(x)$. As a consequence, with $\psi = \phi$, we also have

$$\int_0^T \int_0^\infty B(x)|\tilde{n}(t,x)|dx\ dt \leq C(T, \int |n^0(x)|\ \phi(x)dx) < \infty.$$

Proof of Theorem 3.1. We divide the proof in several steps.

First step. We begin with an existence result for $n^0 \in L^1(\mathbb{R}^+; dx)$. It can be obtained using the Banach–Picard fixed point theorem in the Banach space

$$X = C\big([0,T]; L^1(\mathbb{R}^+; dx)\big), \qquad \|n\|_X = \sup_{0 \leq t \leq T} \|n(t)\|_{L^1(\mathbb{R}^+)},$$

assuming that

$$T\ \|B\|_{L^\infty(\mathbb{R}^+)} \leq \frac{1}{2}. \tag{3.9}$$

With this assumption, we obtain $\tilde{n}$ as the fixed point of the following operator $\mathcal{T}$. Being given $m \in X$, we define $n := \mathcal{T}[m]$ as the solution $n \in X$ of the equation

$$\begin{cases} \frac{\partial}{\partial t}n(t,x) + \frac{\partial}{\partial x}n(t,x) + \lambda_0 n = 0, & t \geq 0,\ x \geq 0, \\[2mm] n(t, x = 0) = \int B(x)m(t,x)dx, \\[2mm] n(t = 0, x) = n^0(x). \end{cases}$$

Therefore, for two functions m_1, m_2, and $n_1 = \mathcal{T}[m_1]$, $n_2 = \mathcal{T}[m_2]$ and $n = n_2 - n_1$, $m = m_2 - m_1$, we have

$$\begin{cases} \frac{\partial}{\partial t}n(t,x) + \frac{\partial}{\partial x}n(t,x) + \lambda_0 n(t,x) = 0, \\[2mm] n(t, x = 0) = \int B(x)m(t,x)dx, \\[2mm] n(t = 0, x) = 0, \end{cases}$$

and thus (see the Appendix 6.1)

$$\begin{cases} \frac{\partial}{\partial t}|n(t,x)| + \frac{\partial}{\partial x}|n(t,x)| + \lambda_0|n(t,x)| = 0, \\[2mm] |n(t, x = 0)| = |\int B(x)m(t,x)dx|, \\[2mm] |n(t = 0, x)| = 0. \end{cases}$$

From this we deduce, after time integration,

$$\|n(t)\|_{L^1(\mathbb{R}^+)} \le \int_0^t |n(s, x = 0)| ds$$

$$= \int_0^t \left| \int B(x) m(s, x) dx \right| ds$$

$$\le t \|B\|_{L^\infty(\mathbb{R}^+)} \sup_{0 \le s \le t} \|m(s)\|_{L^1(\mathbb{R}^+)}.$$

Taking the *sup* on $0 \le t \le T$, we have proved that

$$\|n_1 - n_2\|_X \le \frac{\|B\|_{L^\infty(\mathbb{R}^+)}}{2} \|m_1 - m_2\|_X,$$

and this means that the operator $\mathcal{T}$ is a strict contraction in the Banach space X which proves the existence of a unique fixed point. As usual we can iterate the operator on $[T, 2T]$, $[2T, 3T]$, ..., since the condition on T does not depend on the iteration. With this iteration process, we have built a solution in $C(\mathbb{R}^+; L^1(\mathbb{R}^+))$.

Second step. The comparison principle (ii) follows from the construction of the Banach-Picard fixed point. To show this, consider two initial data n_1^0, n_2^0 and denote by $\mathcal{T}_1$, $\mathcal{T}_2$ the corresponding operators of step 1. If $n_1^0 \le n_2^0$, then for all m we have $\mathcal{T}_1[m] \le \mathcal{T}_2[m]$ and thus the fixed point itself (recall the Picard iteration process) satisfies $n_1 \le n_2$.

As a direct consequence the maximum principle holds true because $\pm C_0 N(x)$ are solutions and can be used in (ii).

Third step. Consider now $n^0 \in L^1(\mathbb{R}^+; \phi(x) dx)$. By density (recall that ϕ is bounded) we can find a sequence $n_k^0 \in L^1(\mathbb{R}^+; dx)$ such that $n_k^0 \xrightarrow[k \to \infty]{} n^0$ in $L^1(\mathbb{R}^+; \phi(x) dx)$. We denote by $\tilde{n}_k(t, x)$ the corresponding solution to (3.8). Combining (3.8) with the dual equation (3.4), we compute for the solution $\tilde{n} = \tilde{n}_k - \tilde{n}_p$,

$$\frac{\partial}{\partial t} (\tilde{n}(t, x) \phi(x)) + \frac{\partial}{\partial x} (\tilde{n}(t, x) \phi(x)) = -\phi(0) B(x) \tilde{n}(t, x). \qquad (3.10)$$

This implies (see the Appendix 6.1 again) the identity

$$\frac{\partial}{\partial t} (|\tilde{n}(t, x)| \phi(x)) + \frac{\partial}{\partial x} (|\tilde{n}(t, x)| \phi(x)) = -\phi(0) B(x) |\tilde{n}(t, x)|,$$

and after integration in x we deduce that

$$\frac{d}{dt} \int |\tilde{n}(t, x)| \phi(x) dx = -\phi(0) \int B(x) |\tilde{n}(t, x)| dx + \phi(0) \left| \int B(x) \tilde{n}(t, x) dx \right| \le 0.$$

Therefore, after time integration,

$$\int |\tilde{n}_k(t) - \tilde{n}_p(t)| \phi dx \le \int |n_k^0 - n_p^0| \phi dx, \qquad (3.11)$$

and thus $(\widetilde{n}_k)$ is a Cauchy sequence of $C\big(\mathbb{R}^+;L^1(\mathbb{R}^+;\phi(x)dx)\big)$. Notice also the uniform bound $|\widetilde{n}_k(t,x)|\leq C_0N(x)$. Therefore it converges in $C\big(\mathbb{R}^+;L^1(\mathbb{R}^+;\phi(x)dx)\big)$, and weakly in $L^\infty(\mathbb{R}^+\times\mathbb{R}^+)$ to a solution in the distribution sense to (3.8).

It is unique because the contraction principle (3.11) shows that two possible solutions coincide on the support of ϕ, thus on the support of B. Then we arrive at two transport equations with the same boundary condition at $x=0$, with the same initial data so these two solutions are equal.

Hence we have proved the well-posedness in $C\big(\mathbb{R}^+;L^1(\mathbb{R}^+;\phi(x)dx)\big)$ and the contraction principle in (iii).

Fourth step. We also deduce from (3.10), after integration in x, that

$$\frac{d}{dt}\int\widetilde{n}(t,x)\phi(x)dx=-\phi(0)\int B(x)\widetilde{n}(t,x)dx+\phi(0)\widetilde{n}(t,0)=0.$$

Thus we have recovered the conservation law in (iii). □

3.4 Regularity of solutions

Later on, we will need some regularity results that we state now. We restrict ourselves to the uniform bounds but any L^p bound on the derivative could be proved as well using the GRE property on the time derivative as we explain it now.

Theorem 3.2. *Assume (3.2) and that the Lipschitz continuous initial data satisfies* $|n^0(x)|\leq C_0N(x)$, $|\frac{\partial}{\partial x}n^0(x)|\leq C_1N(x)$ *and* $n^0(0)=\int B(x)n^0(x)dx$. *Then the solution to (3.8) satisfies*

$$|\frac{\partial}{\partial t}\widetilde{n}(t,x)|\leq(C_1+\lambda_0)N(x)\qquad\forall t\geq0,\ x\geq0, \tag{3.12}$$

$$|\frac{\partial}{\partial x}\widetilde{n}(t,x)|\leq(C_1+\lambda_0+\lambda_0C_0)N(x)\qquad\forall t\geq0,\ x\geq0. \tag{3.13}$$

Proof. We differentiate in t the equation on $\widetilde{n}$ and, setting $q=\frac{\partial}{\partial t}\widetilde{n}$ we see that $q(t,x)$ satisfies also the renewal equation (3.8). Hence we can apply the maximum principle to q (point (i) of Theorem 3.1) and obtain (3.12) noticing that it is true initially because, using equation (3.8) at time $t=0$,

$$|q(t=0,x)|=|\frac{\partial}{\partial t}\widetilde{n}^0|=|\frac{\partial}{\partial x}\widetilde{n}^0+\lambda_0\widetilde{n}^0|\leq(C_1+\lambda_0)N(x),\qquad x>0.$$

On the other hand there is no time jump at $t=x=0$ because we have assumed $n^0(0)=\int B(x)n^0(x)dx$, and thus (3.12) holds up to $x=0$.

The second estimate (3.13) follows because, using again equation (3.8), we have

$$|\frac{\partial}{\partial x}\widetilde{n}(t,x)|=|\frac{\partial}{\partial t}\widetilde{n}(t,x)+\lambda_0\widetilde{n}(t,x)|\leq(C_1+\lambda_0)N(x)+\lambda_0C_0N(x).\qquad□$$

If we do not assume the compatibility condition $n^0(0) = \int B(x)n^0(x)dx$, then we cannot use the equality $\frac{\partial}{\partial t}\tilde{n}^0 = -\frac{\partial}{\partial x}\tilde{n}^0 - \lambda_0\tilde{n}^0$ which fails at $x = 0$ because n^0 has a discontinuity at $x = 0$. In other words $|\frac{\partial}{\partial x}n^0(x)$ has a Dirac mass at $x = 0$ and we cannot use the comparison principle.

3.5 Generalized relative entropy

The contraction principle, (iii) of Theorem 3.1, is a special case of a much more general inequality. It applies to several other equations and all convex functions can be handled rather than the mere absolute value. We present it here and refer to Section 6.4 for a more general setting. It turns out that we could derive here several properties by explicit formulas, but the deep reason they hold true is the generalized relative entropy property. Many other models share several properties with the renewal equation and that is because they also satisfy the GRE property.

Theorem 3.3. *Under the assumptions of Theorem 3.1, then*

(i) *for all convex functions $H : \mathbb{R}^+ \to \mathbb{R}^+$ with $H(0) = 0$, and for all $t > 0$,*

$$\int_0^\infty \phi(x)N(x)H\Big(\frac{\tilde{n}(t,x)}{N(x)}\Big)dx \le \int_0^\infty \phi(x)N(x)H\Big(\frac{n^0(x)}{N(x)}\Big)dx,$$

(ii) *for the probability measure $d\mu(x) = B(x)\frac{N(x)}{N(0)}dx$, and for all convex functions H,*

$$\int_0^\infty \left[\int_0^\infty H\Big(\frac{\tilde{n}(t,x)}{N(x)}\Big)d\mu(x) - H\Big(\int_0^\infty \frac{\tilde{n}(t,x)}{N(x)}d\mu(x)\Big)\right]dt$$
$$\le \int_0^\infty \phi(x)N(x)H\Big(\frac{n^0(x)}{N(x)}\Big)dx.$$

Proof. We use that

$$\frac{\partial}{\partial t}\frac{\tilde{n}(t,x)}{N(x)} + \frac{\partial}{\partial x}\frac{\tilde{n}(t,x)}{N(x)} = 0,$$

so that

$$\frac{\partial}{\partial t}H\Big(\frac{\tilde{n}(t,x)}{N(x)}\Big) + \frac{\partial}{\partial x}H\Big(\frac{\tilde{n}(t,x)}{N(x)}\Big) = 0,$$

and finally

$$\frac{\partial}{\partial t}\Big[\phi(x)N(x)H\Big(\frac{\tilde{n}(t,x)}{N(x)}\Big)\Big] + \frac{\partial}{\partial x}\Big[\phi(x)N(x)H\Big(\frac{\tilde{n}(t,x)}{N(x)}\Big)\Big]$$
$$= -\phi(0)B(x)N(x)H\Big(\frac{\tilde{n}(t,x)}{N(x)}\Big).$$

After integration in $x \in \mathbb{R}^+$ we find, still denoting by $d\mu(x) = B(x)\frac{N(x)}{N(0)}dx$ (a probability measure in view of (3.6)),

$$
\frac{d}{dt}\int \phi(x)N(x)H\Big(\frac{\widetilde{n}(t,x)}{N(x)}\Big)dx
$$

$$
= -\phi(0)N(0)\int H\Big(\frac{\widetilde{n}(t,x)}{N(x)}\Big)d\mu(x) + \phi(0)N(0)H\Big(\frac{\widetilde{n}(t,0)}{N(0)}\Big)
$$

$$
= \phi(0)N(0)\Big[-\int H\Big(\frac{\widetilde{n}(t,x)}{N(x)}\Big)d\mu(x) + H\Big(\int \frac{\widetilde{n}(t,x)}{N(x)}d\mu(x)\Big)\Big] \le 0,
$$

for all convex functions H with $H(0) = 0$. The statements (i) and (ii) follow from this inequality. $\qquad\square$

3.6   Long time asymptotic: entropy method

In practice one observes the *Stable Age Distribution*, i.e., the long time limit of $\widetilde{n}$ which is expected to be proportional to the steady state N given by equation (3.3). In this section, we prove a general statement without a rate. Exponential rate of convergence is treated afterwards in Section 3.7 where we give a simple method using a restrictive hypothesis which is improved in Section 3.9.1.

We recall the notation $\widetilde{n} = ne^{-\lambda_0 t}$ which satisfies equation (3.8).

Theorem 3.4. *Under assumptions (3.2) and $|n^0(x)| \le CN(x)$, the solution to (3.1) given by Theorem 3.1 satisfies*

$$
\int_0^\infty |\widetilde{n}(t,x) - m^0 N(x)|\phi(x)dx \downarrow 0 \qquad as \ \ t \to \infty, \tag{3.14}
$$

with $m^0 = \int n^0(x)\phi(x)dx$ (a conserved quantity).

Proof. First step. We can always regularize the initial data n^0 in $L^1(\phi(x)dx$ so as to satisfy the assumptions of Theorem 3.2. Call n^0_ε the corresponding new initial data, then, using the contraction principle we have

$$
\int_0^\infty |n - n_\varepsilon|(t,x)\phi(x)dx \le \int_0^\infty |n^0 - n^0_\varepsilon|\phi(x)dx := r_\varepsilon \to 0,
$$

and also $|m^0 - m^0_\varepsilon| \le r_e$. Therefore it is enough to prove the result for the regularized initial data. Indeed

$$
\int_0^\infty |\widetilde{n}(t,x) - m^0 N(x)|\phi(x)dx \le 2r_\varepsilon + \int_0^\infty |\widetilde{n}_\varepsilon(t,x) - m^0_\varepsilon N(x)|\phi(x)dx,
$$

and thus $\lim_{t\to\infty}\int_0^\infty |\widetilde{n}(t,x) - m^0 N(x)|\phi(x)dx \le 2r_\varepsilon$ for all $\varepsilon > 0$, which proves that the limit vanishes because $r_\varepsilon \to 0$.

Second step. In the regularized case, we set $h(t, x) = \widetilde{n}(t, x) - m^0 N(x)$, which is also a solution to equation (3.8), satisfies $|h(t, x)| \leq C_0 N(x)$ and $\int h(t, x)\phi(x)dx = 0$ (using the conservation property (iii) of Theorem 3.1). We prove that such a solution vanishes over a long time. Notice that, by the GRE property, we have

$$\int_0^\infty |h(t, x)|\phi(x)dx \downarrow L, \qquad \text{as } t \to \infty.$$

And it remains to show that $L = 0$.

Third step. Always in the regularized case, we define the solution to equation (3.8), $h_k \in C(\mathbb{R}^+; L^1(\mathbb{R}^+; \phi(x)dx))$, by

$$h_k(t, x) = h(t + k, x), \quad \text{and thus } |h_k| \leq C_0 N.$$

Using Theorem ii, for $H(\cdot)$ convex with $H(0) = 0$, we know that

$$\int_0^\infty \Big[\int H\Big(\frac{h_k(t, x)}{N(x)}\Big)d\mu(x) - H\Big(\int_0^\infty \frac{h_k(t, x)}{N(x)}d\mu(x)\Big)\Big] dt := I_k \to 0, \qquad \text{as } k \to \infty.$$

$$(3.15)$$

Indeed, from the very definitions of I_k and h_k, we have

$$I_k = \int_k^\infty \Big[\int H\Big(\frac{h(t, x)}{N(x)}\Big)d\mu(x) - H\Big(\int_0^\infty \frac{h(t, x)}{N(x)}d\mu(x)\Big)\Big] dt$$

and the integrand is nonnegative and integrable.

We also notice that h_k satisfies equation

$$\begin{cases} \frac{\partial}{\partial t} h_k(t, x) + \frac{\partial}{\partial x} h_k(t, x) + \lambda_0 h_k(t, x) = 0, & t \geq 0, \, x \geq 0, \\ h_k(t, x = 0) = \int B(y)h_k(t, y)dy, & \int_0^\infty \phi(x)h_k(t, x)dx = 0. \end{cases}$$

$$(3.16)$$

Next, using the regularity of h (via Theorem 3.2), we may extract a subsequence (still denoted h_k) such that, for all $T > 0$,

$$\begin{cases} h_k \to g \quad \text{in } C([0, T] \times \mathbb{R}^+), \qquad 0 \leq |g| \leq C_0 N(x), \\ \int B(y)h_k(t, y)dy \to \int B(y)g(t, y)dy \quad \text{in } C([0, T]), \\ \int_0^\infty \phi(x)g(t, x)dx = 0, \qquad \int_0^\infty \phi(x)|g(t, x)|dx = L. \end{cases}$$

We pass to the limit in the entropy relation (3.15) and obtain, by convexity in weak limits,

$$\int_0^\infty \int H\Big(\frac{g(t, x)}{N(x)}\Big)d\mu(x)dt \leq \lim \int_0^\infty \int H\Big(\frac{h_k(t, x)}{N(x)}\Big)d\mu(x)dt$$
$$= \int_0^\infty H\Big(\int_0^\infty \frac{g(t, x)}{N(x)}d\mu(x)\Big)dt.$$

$$(3.17)$$

But from the Jensen inequality, the reverse inequality is also true thus showing that

$$\int_0^\infty \int H\Big(\frac{g(t,x)}{N(x)}\Big)d\mu(x)dt \geq \int_0^\infty H\Big(\int_0^\infty \frac{g(t,x)}{N(x)}d\mu(x)\Big)dt.$$

This equality for H strictly convex shows that for almost all $t > 0$, on the support of μ, i.e., that of B,

$$\frac{g(t,x)}{N(x)} = C(t) \quad \text{(independent of } x \in \text{supp B).}$$

Inserting this information in the limit in distribution sense of equation (3.16), namely

$$\begin{cases} \frac{\partial}{\partial t}g(t,x) + \frac{\partial}{\partial x}g(t,x) + \lambda_0\, g(t,x) = 0, \quad t \geq 0,\ x \geq 0, \\[2mm] g(t,x=0) = \int B(y)g(t,y)dy, \end{cases}$$

and thus

$$\frac{\partial}{\partial t}\frac{g(t,x)}{N(x)} + \frac{\partial}{\partial x}\frac{g(t,x)}{N(x)} = 0.$$

Therefore $\frac{g(t,x)}{N(x)} \equiv C(t)$ and $C(t)$ is in fact a constant in t;

$$g(t,x) = G^0\, N(x) \qquad \forall x \geq 0.$$

Using that $\int_0^\infty g(t,x)\phi(x) = 0$, we find that $G^0 = 0$. Now we can conclude that the limit L of the second step vanishes because, passing to the limit $k \to \infty$ we have $L = \int_0^\infty |g(t,x)|\phi(x) \to 0$ for $t \to \infty$. $\qquad\square$

3.7   Long time asymptotic: exponential decay

Theorem 3.5. *Under assumption* (3.2), *and*

$$\exists \mu_0 > 0, \quad s.t. \quad B(x) \geq \mu_0 \frac{\phi(x)}{\phi(0)}, \tag{3.18}$$

the solution to (3.1) *satisfies*

$$\int |\tilde{n}(t,x) - m^0 N(x)|\phi(x)dx \leq e^{-\mu_0 t}\int |n^0(x) - m^0 N(x)|\phi(x)dx \tag{3.19}$$

with $m^0 = \int_0^\infty n^0(x)\phi(x)dx$ *a conserved quantity, see Theorem* 3.1 (ii).

The assumption (3.18) is restrictive if we have in mind that B can vanish for $x \approx 0$ and be positive afterwards because $\phi(x) > 0$ on the convex hull of the support of B. But for x large, or close to the end point of the support of B, in general the quantity $\phi(x)$ vanishes faster than B. See Figure 3.1.

There are two proofs of exponential time convergence that do not use (3.18). One is through the Laplace transform, and gives a representation of the solution but not a precise time decay in functional spaces ([171, 143, 100]). The other one, [124], is based on the invariants of the renewal equation (surprisingly there are several conserved quantities like $\int \tilde{n}\phi(x)dx$).

Proof. We define
$$h(t, x) = \tilde{n}(t, x) - mN(x).$$

By linearity it still satisfies the equation
$$\begin{cases} \frac{\partial}{\partial t}h(t, x) + \frac{\partial}{\partial x}h(t, x) + \lambda_0\, h(t, x) = 0, \quad t \geq 0,\ x \geq 0, \\ \\ h(t, x = 0) = \int B(y)h(t, y)dy. \end{cases}$$

Since equation (3.4) is the dual equation, we have again by a simple combination of these two equations,
$$\begin{cases} \frac{\partial}{\partial t}\big[h(t, x)\phi(x)\big] + \frac{\partial}{\partial x}\big[h(t, x)\phi(x)\big] = -\phi(0)B(x)h(t, x), \quad t \geq 0,\ x \geq 0, \\ \\ \phi(0)h(t, x = 0) = \phi(0)\int B(y)h(t, y)dy. \end{cases}$$

Therefore (see Section 6.1)
$$\begin{cases} \frac{\partial}{\partial t}\big[|h(t, x)|\phi(x)\big] + \frac{\partial}{\partial x}\big[|h(t, x)|\phi(x)\big] = -\phi(0)B(x)|h(t, x)|, \quad t \geq 0,\ x \geq 0, \\ \\ \phi(0)|h(t, x = 0)| = \phi(0)|\int B(y)h(t, y)dy|. \end{cases}$$

After integration in x, we obtain since $\int \phi(x)h(t, x)dx = 0$,

$$\begin{aligned} \frac{d}{dt}\int |h(t, x)|\phi(x)dx &= -\phi(0)\int B(x)|h(t, x)|dx + \phi(0)|\int B(x)h(t, x)dx| \\ &= -\phi(0)\int B(x)|h(t, x)|dx + |\int[\phi(0)B(x) - \mu_0\phi(x)]h(t, x)dx| \\ &\leq -\phi(0)\int B(x)|h(t, x)|dx + \int[\phi(0)B(x) - \mu_0\phi(x)]|h(t, x)|dx \\ &= -\mu_0\int |h(t, x)|\phi(x)dx. \end{aligned}$$

We conclude using Gronwall's lemma. □

We conclude this section by an explicit example. We take
$$B(x) = \nu\, e^{-\mu x}, \qquad \nu > \mu,$$

and thus, assumption (3.2) is satisfied. One computes
$$\lambda_0 = \nu - \mu, \quad \phi(x) = \phi(0)e^{-\mu x}, \quad N(x) = \lambda_0 e^{-lbx}.$$

Therefore the assumption (3.18) is satisfied with $\mu_0 = \nu$.

3.8   Extension: death rates

As an extension of the previous section, we consider the renewal equation with a death rate $d(x) \geq 0$,

$$
\begin{cases}
\frac{\partial}{\partial t} n(t,x) + \frac{\partial}{\partial x} n(t,x) + d(x)\, n(t,x) &= 0, \quad t \geq 0, \, x \geq 0, \\[2mm]
n(t, x = 0) &= \int B(y) n(t,y) dy.
\end{cases}
\tag{3.20}
$$

In practice, the death rate $d(x)$ can vary drastically for different species: for trees it is often constant, for fishes it decreases for small values of x (smaller fish are easier targets for predators) and increases for larger x, for mammals it is often increasing and unbounded; Gompertz's law gives a linear growth of $d(x)$ (this is 'aging', the probability of death increases with age) after some age before which it is constant (when death arises only by stochastic events independently of age). More recent theories tend to show that after some age $d(x)$ becomes again constant.

We define

$$
D(x) = \int_0^x d(y) dy.
$$

Then the equation on $m(t,x) = n(t,x) e^{D(x)}$ is

$$
\begin{cases}
\frac{\partial}{\partial t} m(t,x) + \frac{\partial}{\partial x} m(t,x) &= 0, \quad t \geq 0, \, x \geq 0, \\[2mm]
m(t, x = 0) &= \int B(y) e^{-D(y)} m(t,y) dy.
\end{cases}
\tag{3.21}
$$

Therefore we are in the simpler situation studied so far.

Exercise. Assume that

$$
B \geq 0, \qquad B \in L^\infty(\mathbb{R}^+), \qquad \int B(x) e^{-D(x)} dx > 1.
\tag{3.22}
$$

Write the eigenvalue problem (λ, N, ϕ) for equation (3.20) and show that there exists a solution with $\lambda_0 > 0$ and $\int_0^\infty N = 1$. Extend the existence theory and the generalized relative entropy inequality. Make the relation between the eigenelements of equation (3.20) and those of (3.21).

Exercise. Assume that

$$
B \geq 0, \qquad B \in L^\infty(\mathbb{R}^+), \qquad d(x) \xrightarrow[x \to \infty]{} \infty.
$$

Show that there exist eigenelements (λ_0, N, ϕ), $\lambda_0 \in \mathbb{R}$ (no sign condition in this case) with $\int_0^\infty N = 1$.

At this level we would like to remark that there is no reason to look for a positive eigenvalue λ_0. Species can also go extinct in certain conditions, a situation which corresponds to $\lambda_0 < 0$. However in the case $d \equiv 0$, the condition $\lambda_0 > 0$

is necessary otherwise the equation on N only has growing exponential solutions which do not really make sense (in terms of biological interpretation), and which would require very specific assumptions on B such as $B(x) = 0$ for $x \geq x^\sharp$.

We also refer to Section 3.9.1 for a more specific example with a death rate.

3.9 More realistic models around the renewal equation

Many variants of the renewal equations have been used and studied in various areas of biology. We give here several examples and explain roughly the modeling behind them. They are often nonlinear and we recall that such models were already mentioned in Section 1.5.2 with the Kermack–McKendrick model in epidemiology.

3.9.1 Renewal equation for cell division cycle (one phase)

This section presents some improvement of the arguments in Section 3.7 to obtain exponential time decay. We show that it is possible to prove an explicit exponential rate of convergence, in the natural norm, for situations more general than the mere assumption (3.18). We illustrate the idea on a model for cell division cycle with a single phase, see Section 3.9.5 for a motivation. This is a special case of an age structured equation written as

$$
\begin{cases}
\frac{\partial}{\partial t} n(t, x) + \frac{\partial}{\partial x} n(t, x) + k(x)\, n(t, x) = 0, & t \geq 0,\ x \geq 0, \\[2mm]
n(t, x = 0) = 2 \int k(y) n(t, y) dy, \\[2mm]
n(t = 0, x) = n^0(x),
\end{cases}
\tag{3.23}
$$

where the function $k(\cdot)$ can be interpreted as a mitosis rate. When cells undergo mitosis, they are withdrawn from the balance equation at age x with a rate $k(x)$ and create two daughter cells at age $x = 0$ with the same rate.

For our theoretical study, we take for simplicity the mitosis rate

$$
k(x) = \beta\, \mathbf{I}_{\{x \geq x_*\}}, \qquad \beta > 0, \quad x_* > 0.
\tag{3.24}
$$

We recall here the definitions of the eigenelements, and readily check that they exist here because condition (3.22) is fulfilled,

$$
\begin{cases}
\frac{\partial}{\partial x} N(x) + (k(x) + \lambda_0)\, N(x) = 0, & x \geq 0, \\[2mm]
N(x = 0) = 2 \int k(y) N(y) dy, & N > 0, \quad \int N(x) dx = 1,
\end{cases}
$$

$$
\begin{cases}
\frac{\partial}{\partial x} \phi(x) - (k(x) + \lambda_0)\, \phi(x) = -2\phi(0) k(x), & x \geq 0, \\[2mm]
\phi \geq 0, \quad \int \phi(x) N(x) dx = 1.
\end{cases}
$$

Because k vanishes close to 0, assumption (3.18) does not hold here. Nevertheless we can obtain an exponential rate of convergence.

Theorem 3.6. *We assume* (3.24) *and*

$$e^{-\beta x_*/2} > \frac{3}{4}, \tag{3.25}$$

then, for some positive function $\bar{\phi}$ and μ_0 given in (3.27), *we have*

$$\int |\tilde{n}(t,x) - m^0 N(x)| \bar{\phi}(x) dx \le e^{-\mu_0 t} \int |n^0(x) - m^0 N(x)| \bar{\phi}(x) dx, \tag{3.26}$$

with $m^0 = \int n^0 \phi$ and some $\mu_0 > 0$ given in the proof below.

Proof. First step. We compute the eigenvalue λ_0. The basic functions N, $Q = N\phi$ are given in this case by

$$N(x) = \begin{cases} e^{-\lambda_0 x} & for \ x \le x_*, \\ e^{-(\lambda_0 + \beta)x} \ e^{\beta x_*} & for \ x \ge x_*, \end{cases}$$

$$Q(x) = \begin{cases} 1 & for \ x \le x_*, \\ 2\frac{\beta}{\lambda_0 + \beta} N(x) & for \ x \ge x_*. \end{cases}$$

Especially, λ_0 is defined by $2\int k(x)N(x)dx = 1$ which gives

$$\lambda_0 + \beta = 2\beta e^{-\lambda_0 \ x_*}, \tag{3.27}$$

an equation which always has a solution $0 < \lambda_0 < \beta$.

Second step. We consider a function $c(x)$ to be chosen later (that replaces $2k(x)$ in the right-hand side of the equation defining ϕ) such that

$$c(x) \ge 0, \qquad \int c(x)N(x)dx = 1. \tag{3.28}$$

And we set

$$\bar{Q}(x) = 1 - \int_0^x c(y)N(y)dy, \qquad \bar{Q} = N\bar{\phi}. \tag{3.29}$$

As in the proof of Theorem 3.5, we consider the function $h = n(t,x)e^{-\lambda_0 t} - m^0 N(x)$, which still satisfies equation (3.23) (with a different Cauchy data) and $\int h(t,x)\phi(x)dx = 0$. We have by a simple combination of the equation on h and the equation

$$\frac{\partial}{\partial x}\bar{\phi}(x) - k(x)\bar{\phi}(x) = -c(x),$$

$$\begin{aligned} \frac{d}{dt}\int |h(t,x)|\bar{\phi}(x)dx &= -\int |h(t,x)|c(x)dx + \int h(t,x)2k(x)dx \\ &= -\int |h(t,x)|c(x)dx + \int h(t,x)[2k(x) - \mu\phi(x)]dx \\ &\le -\int |h(t,x)|c(x)dx + \int |h(t,x)||2k(x) - \mu\phi(x)|dx. \end{aligned}$$

In order to choose $c(x) = \mu_0 \bar{\phi}(x) + |2k(x) - \mu\phi(x)|$, for some $\mu_0 > 0$, and since c is only constrained via (3.28), it is enough to be sure that

$$\int |2k(x) - \mu\phi(x)|N(x)dx < 1.$$

But, assuming $2k(x) > \mu\phi(x)$ for $x > x_*$, we have

$$\int_0^{x_*} \mu\phi(x) + \int_{x_*}^{\infty} [2k(x) - \mu\phi(x)]N(x)dx = \frac{\beta}{\lambda_0}(e^{-\lambda_0 x_*} - 1) + 1 - 2\frac{\beta^2}{(\lambda_0 + \beta)^2}e^{-\lambda_0 x_*},$$

and this quantity is less than 1 iff $\beta < 2\lambda_0$, which is satisfied, in view of the definition (3.27) of λ_0 iff

$$\frac{\beta}{2} + \beta < 2\beta e^{-\frac{\beta}{2}x_*},$$

which itself is equivalent to (3.25) and is compatible with the sign condition used above for $x > x_*$. $\qquad\square$

3.9.2   Cell division cycle: numerical results

We now present (Figures 3.2 and 3.3) some numerical results in order to illustrate the time decay result of Theorem ii.

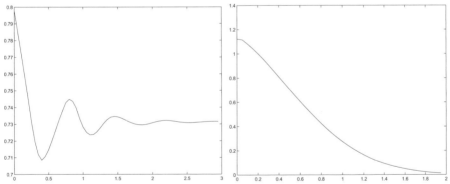

Figure 3.2: SOLUTION OF THE RENEWAL EQUATION IN CASE 1 (PEAKED RENEWAL KERNEL). LEFT: THE TOTAL POPULATION DENSITY AS A FUNCTION OF TIME. RIGHT: THE STEADY STATE $N(x)$.

We consider the equation (3.20) with a death rate $d(x) = 2. * x/x_M + k(x)$ where we have fixed an adimensionalisation parameter x_M, the birth rate $b(x) = 2 * k(x)$ (see the choice of $k(x)$ below) is motivated by a simple case of cell division cycle (Section 3.9.1). We consider the solution up to the final time $t_{final} = 3 * x_M$. The eigenvalue λ_0 is computed numerically (and thus depends upon the discretization). One can observe that, after renormalization, the solution converges to a steady state but with an initial oscillatory behavior. This kind of behavior has

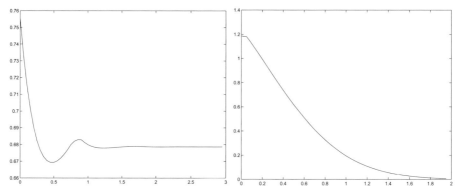

Figure 3.3: Solution of the renewal equation in case 2 (uniform renewal kernel). Left: the total population density as a function of time. Right: the steady state $N(x)$.

been confirmed and matched to experiments on various types of cells in [221], [60] ... etc.

As one can see in the numerical tests below, these oscillations are related to the form of the kernel $k(a)$ used above. We can see that the more 'peaked' is the kernel K, the more oscillations the solution exhibits, thanks to the two cases:

- First case. $k(x) = 4 * \mathbf{1}_{\{x_M/2 \leq x \leq x_M\}}$, see Figure 3.2.
- Second case. $k(x) = 2 * \mathbf{1}_{\{0 \leq x \leq x_M\}}$,   see Figure 3.3.

3.9.3   A renewal system with quiescent/proliferative cells

With the same convention as before, we consider another example of an age structured model. We let $x \geq 0$ denote the age of cells but we now assume two states for cells: the proliferative state with cell density $p(t,x)$ and the quiescent state with cell density $q(t,x)$ (this can be also called 'phase G0' with the terminology of Section 3.9.5), thus extending the model presented in Section 3.9.1. The balance equations for the two compartments are

$$\begin{cases} \frac{\partial}{\partial t}p(t,x) + \frac{\partial}{\partial x}p(t,x) + [B(x) + d(x) + \sigma_1(x)]\,p(t,x) & = \sigma_2(x)\,q(t,x), \\[2mm] \frac{\partial}{\partial t}q(t,x) + \frac{\partial}{\partial x}q(t,x) + \sigma_2(x)\,q(t,x) & = \sigma_1(x)\,p(t,x), \\[2mm] p(t,x=0) = 2\int B(y)p(t,y)dy, \qquad q(t,x=0) = 0, \\[2mm] p(t=0,x) = p^0(x), \qquad q(t=0,x) = q^0(x). \end{cases} \tag{3.30}$$

Here σ_i are the transition rates from one state to the other and $B(x)$ is the mitosis rate for cells in the proliferative state, since it is assumed that in the quiescent state there is no birth.

This model has been studied in [10] and in particular its long time behavior. It is however easy to study it along with the general theory developed here. There are eigenelements (λ_0, P, Q), with $\lambda_0 > 0$, $P(x > 0)$, $Q(x > 0)$ under fairly general conditions; the formulas are explicit but long to write. Then one has a Generalized Relative Entropy inequality, following Sections 3.5, 6.3 and 6.4, that explains why $(pe^{-\lambda_0 t}, qe^{-\lambda_0 t})$ converges to a multiple of (P, Q), again in spaces with the correct weights given by the adjoint equation.

3.9.4 The renewal equation with diffusion

We now consider the renewal equation with diffusion and show that the Generalized Relative Entropy method developed in Section 3.6 applies directly.

The equation with diffusion is written

$$
\begin{cases}
\frac{\partial}{\partial t} n(t, x) + \frac{\partial}{\partial x} n(t, x) + d(x)\, n(t, x) = \frac{\partial}{\partial x}[\nu(x)\frac{\partial}{\partial x} n(x)], & t \geq 0,\ x \geq 0, \\[2mm]
n(t, x = 0) - \nu(0)\frac{\partial}{\partial x} n(x = 0) = \int B(y)n(t, y)dy, \\[2mm]
n(t = 0, x) = n^0(x).
\end{cases}
\tag{3.31}
$$

Our notation follows those of Section 3.6; λ_0, N and ϕ are the first eigenelements of the stationary problem and dual problem

$$
\begin{cases}
\frac{\partial}{\partial x} N(x) + (d(x) + \lambda_0)\, N(x) = \frac{\partial}{\partial x}[\nu(x)\frac{\partial}{\partial x} N(x)], & x \geq 0, \\[2mm]
N(x = 0) - \nu(0)\frac{\partial}{\partial x} N(x = 0) = \int B(y)N(y)dy, & \int N(x)dx = 1,
\end{cases}
\tag{3.32}
$$

$$
\begin{cases}
-\frac{\partial}{\partial x} \phi(x) + (d(x) + \lambda_0)\, \phi(x) = \frac{\partial}{\partial x}[\nu(x)\frac{\partial}{\partial x} \phi(x)] + b(x)\phi(0), & x \geq 0, \\[2mm]
\nu(0)\frac{\partial}{\partial x} \phi(0) = 0, & \phi \geq 0, \quad \int N(x)\phi(x)dx = 1.
\end{cases}
\tag{3.33}
$$

Theorem 3.7. *The entropy equalities hold, for all convex functions $H(\cdot)$ and with the notation $\widetilde{n} = e^{-\lambda_0 t} n$,*

$$
\frac{\partial}{\partial t} \int N(x)\phi(x)H\Big(\frac{\widetilde{n}(t, x)}{N(x)}\Big)dx = -D_{\text{diff}}^H\big(\widetilde{n}(t)\big) - D_{\text{ren}}^H\big(\widetilde{n}(t)\big) \leq 0,
$$

where the entropy dissipation due to diffusion and to the renewal terms are

$$
D_{\text{diff}}^H\big(\widetilde{n}(t)\big) = \int H''\Big(\frac{\widetilde{n}(t, x)}{N(x)}\Big)\, \nu(x)N(x)\phi(x)\Big(\frac{\partial}{\partial x}\frac{\widetilde{n}(t, x)}{N(x)}\Big)^2 dx \geq 0,
$$

$$
D_{\text{ren}}^H\big(\widetilde{n}(t)\big) = \phi(0)\int \Big[H\Big(\frac{\widetilde{n}(t, x)}{N(x)}\Big) - H\Big(\frac{\widetilde{n}(t, 0)}{N(0)}\Big)
$$

$$
- H'\Big(\frac{\widetilde{n}(t, 0)}{N(0)}\Big)\Big(\frac{\widetilde{n}(t, x)}{N(x)} - \frac{\widetilde{n}(t, 0)}{N(0)}\Big)\Big]b(x)N(x)dx \geq 0.
$$

Proof. We deduce from the equation on $N(x)$ the additional equality

$$\frac{\partial}{\partial x}\frac{1}{N(x)} - (d(x)+\lambda_0)\frac{1}{N(x)} = \frac{\partial}{\partial x}[\nu(x)\frac{\partial}{\partial x}\frac{1}{N(x)}] - 2\nu(x)N(x)(\frac{\partial}{\partial x}\frac{1}{N(x)})^2. \quad (3.34)$$

Combining equations (3.31) and (3.34) gives us

$$\frac{\partial}{\partial t}\frac{\tilde{n}(t,x)}{N(x)} + \frac{\partial}{\partial x}\frac{\tilde{n}(t,x)}{N(x)} = \frac{\partial}{\partial x}[\nu(x)\frac{\partial}{\partial x}\frac{\tilde{n}(t,x)}{N(x)}]$$
$$-2\nu(x)\frac{\partial}{\partial x}\tilde{n}(t,x)\frac{\partial}{\partial x}\frac{1}{N(x)} - 2\nu(x)\tilde{n}(t,x)N(x)(\frac{\partial}{\partial x}\frac{1}{N(x)})^2$$
$$= \frac{\partial}{\partial x}[\nu(x)\frac{\partial}{\partial x}\frac{\tilde{n}(t,x)}{N(x)}] - 2\nu(x)N(x)\frac{\partial}{\partial x}\frac{1}{N(x)}\frac{\partial}{\partial x}\frac{\tilde{n}(t,x)}{N(x)}.$$

For any function $H(\cdot)$ we deduce an equality similar to the above, namely

$$\frac{\partial}{\partial t}H\left(\frac{\tilde{n}(t,x)}{N(x)}\right) + \frac{\partial}{\partial x}H\left(\frac{\tilde{n}(t,x)}{N(x)}\right) = \frac{\partial}{\partial x}[\nu\frac{\partial}{\partial x}H\left(\frac{\tilde{n}(t,x)}{N(x)}\right)] - H''\left(\frac{\tilde{n}(t,x)}{N(x)}\right)\nu(x)\left(\frac{\partial}{\partial x}\frac{\tilde{n}(t,x)}{N(x)}\right)^2$$
$$-2\nu(x)N(x)\frac{\partial}{\partial x}\frac{1}{N(x)}\frac{\partial}{\partial x}H\left(\frac{\tilde{n}(t,x)}{N(x)}\right).$$

And it remains to combine it with the equation for N to 'undo' the conservative form on $\frac{\tilde{n}(t,x)}{N(x)}$ and arrive at

$$\frac{\partial}{\partial t}N(x)H\left(\frac{\tilde{n}(t,x)}{N(x)}\right) + \frac{\partial}{\partial x}N(x)H\left(\frac{\tilde{n}(t,x)}{N(x)}\right) + d(x)N(x)H\left(\frac{\tilde{n}(t,x)}{N(x)}\right)$$
$$= \frac{\partial}{\partial x}[\nu\frac{\partial}{\partial x}N(x)H\left(\frac{\tilde{n}(t,x)}{N(x)}\right)] - H''\left(\frac{\tilde{n}(t,x)}{N(x)}\right)\nu N(x)\left(\frac{\partial}{\partial x}\frac{\tilde{n}(t,x)}{N(x)}\right)^2.$$

We are now close to the entropy inequality for this problem. Using the dual solution ϕ gives the entropy form

$$\frac{\partial}{\partial t}\int N(x)\phi(x)H\left(\frac{\tilde{n}(t,x)}{N(x)}\right)dx = -\int H''\left(\frac{\tilde{n}(t,x)}{N(x)}\right)\nu(x)N(x)\phi(x)\left(\frac{\partial}{\partial x}\frac{\tilde{n}(t,x)}{N(x)}\right)^2 dx$$
$$+\phi(0)\left[N(0)H\left(\frac{\tilde{n}(t,0)}{N(0)}\right) - \int H\left(\frac{\tilde{n}(t,x)}{N(t,x)}\right)b(x)N(x)dx\right.$$
$$-\nu(0)\left[H\left(\frac{\tilde{n}(t,0)}{N(0)}\right)\frac{\partial N(0)}{\partial x} + N(0)H'\left(\frac{\tilde{n}(t,0)}{N(0)}\right)\frac{\partial}{\partial x}\frac{\tilde{n}(0)}{\partial N(0)}\right].$$
$$(3.35)$$

The first line represents the entropy dissipation by diffusion D_{diff}^H, and the boundary terms can be treated as follows. Combining the boundary terms on $n(t)$ and N we first deduce that

$$\nu(0)N(0)\frac{\partial}{\partial x}\frac{\tilde{n}(0)}{\partial N(0)} = \frac{\tilde{n}(t,0)}{N(0)}\int b(x)N(x)dx - \int b(x)\tilde{n}(t,x)dx,$$

and we arrive, in (3.35), at the term

$$\phi(0)\left[H\left(\frac{\tilde{n}(0)}{N(0)}\right)[N(0) - \nu(0)\frac{\partial}{\partial x}N(0)]\right.$$
$$+\int_{\mathbb{R}^+}\left[-H\left(\frac{\tilde{n}(t,x)}{N(t,x)}\right) + H'\left(\frac{\tilde{n}(0)}{N(0)}\right)\left(\frac{\tilde{n}(t,x)}{N(t,x)} - \frac{\tilde{n}(t,0)}{N(0)}\right)\right]b(x)N(x)dx\right],$$

which is the term $-D_{\text{ren}}^H$ and which sign results from convexity of $H(\cdot)$. $\square$

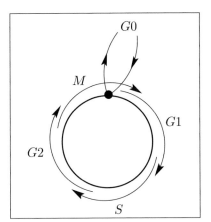

Figure 3.4: PRINCIPLE OF THE CELL DIVISION CYCLE AND ITS PHASES.

3.9.5 A system of renewal equations for cell division cycle

The cell division (mitosis) is the result of a full cycle that a cell should undergo successfully ([150], Ch. 13). It is usually accepted that the cycle consists in four phases as depicted in Figure 3.4, (i) a growth phase, denoted G1 (but G stands for gap), where approximately the cell doubles its size (this is at least well reported for yeast), (ii) a synthesis phase, denoted by S, where the DNA is duplicated, (iii) a rest phase, denoted by G2, a usual interpretation is that it is used to check and repair the errors in S phase, (iv) the mitosis itself, phase M, where the two DNA folds separate (anaphase) and the cell divides finally. It was discovered in the early 1980s that the cell division cycle progression is related to variations of the concentration of certain proteins called cyclins (cyclin B in phase M, cyclin A in phase S, cyclin C to F in phase G_1) and thus we wish to keep this notion in the mathematical model, see [206]. Also many of the cells stay at rest in a quiescent phase called G0, typically, skin cells are constantly in the cycle but endothelial cells are known to stay at rest in phase G0 (also called quiescent state) and may be activated by Vascular Endothelial Growth Factors when angiogenesis occurs (see Section 5.5.6). On the other hand, the cell division cycle duration is also extremely variable, from several minutes to several days, depending on the cells [150], [120].

To fit with the above description, the cell population model uses the density of cells in the phase i, denoted by $n_i(t, x)$, at time t and structured with age x in the cell, and say $i = 0$ for the rest phase G0, and $i = 1, 2 \ldots, I$ with $I = 4$ in the above scenario. We call $K_{i \to i+1}(x) \geq 0$ the transition rate from phase i to phase $i + 1$ depending on the age x in the phase i ($K_{I \to I+1}$ stands sometimes for $K_{I \to 1}$ in order to simplify some notation). As a simple model we can have in mind

$$K_{i \to i+1}(x) = k_i \mathbf{1}_{\{x \geq x_i\}}, \qquad k_i > 0, \tag{3.36}$$

a constant rate after some age x_i is attained. These transition rates could be controlled by drugs or by the circadian clock, thus leading to a variant

$$K_{i \to i+1}(t, x) = k_i(t) \mathbf{I}_{\{x \geq x_i\}}, \tag{3.37}$$

then $k_i(t)$ is switch activated periodically at certain times in the circadian control, by a therapy in case of therapeutic control. In the case when $k_i(t)$ are periodic, one can apply Floquet theory in order to extend the results of this chapter along the lines of Section 6.3.2.

Then the evolution of the density in the phase is modeled via a renewal equation as studied in Section 3.1. We denote by $d_i > 0$ the death rate in the phase i : as mentioned earlier, it could also be controlled by drugs and incorporate a circadian rhythm; $v_i(x) > 0$ is the evolution speed in the phase which could also be controlled by some cyclin level for instance. Hence, for $1 \leq i \leq I$, $t \geq 0$ and $a \geq 0$, we have

$$\begin{cases} \frac{\partial}{\partial t} n_i(t, x) + \frac{\partial}{\partial x}[v_i(x) n_i(t, x)] + [d_i(x) + K_{i \to i+1}(x)] n_i(t, x) = 0, \\[2mm] v_i(0) n_i(t, x = 0) = \int_{x' \geq 0} K_{i-1 \to i}(x') \, n_{i-1}(t, x') \, dx', \qquad 2 \leq i \leq I, \quad (3.38) \\[2mm] v_1(0) n_1(t, x = 0) = 2 \tau_M \int_{x' \geq 0} K_{I \to 1}(x') \, n_I(t, x') \, dx', \end{cases}$$

where the factor 2 expresses the doubling of cell number after mitosis phase M and $0 < \tau_M \leq 1$ is the rate of cells which continue the cycle after mitosis. For the sake of simplicity we have not described the phase $G0$ which receives the extra cells after mitosis and possibly can introduce new cells in the phase $G1$. Of course our model is completed by a set of Cauchy data

$$n_i(t = 0, x) = n_i^0(x) \geq 0, \qquad \forall i = 1, \ldots, I, \quad \forall x > 0. \tag{3.39}$$

This model thus retains some aspects of Rotenberg's [205] model (see Section 3.9.6) with a discrete set of maturation states (μ in (3.40) corresponding to i in (3.38)) while keeping the main feature that enough phase progression is needed for transition to the next phase. Several variants exist and in particular age is not always the best variable. For instance during the S phase, one can measure the DNA content that should be doubled at the end ([20]). Also the content in cyclins is a better structuring parameter than age ([22]).

One can develop a theory similar to that of Section 3.2 for the existence of eigenelements and the long time convergence to the first eigenvector. Namely we have (assuming that all coefficients are bounded and v_i does not vanish)

Theorem 3.8. *There is a unique first eigenelement* ($\lambda_0 > 0$, $N_i > 0$, $\phi_i \geq 0$) *(normalized as usual) under the condition*

$$2 \tau_M \prod_{i=1}^{I} \int_{x \geq 0} K_{i \to i+1}(x) e^{-\int_0^x \frac{d_i + K_{i \to i+1}}{v_i}} \, dx > 1.$$

This condition, that replaces (3.2), is obtained by an exact integration of the steady state equation and following the arguments of Section 3.2.

3.9.6 Maturation structure

Somewhat related but more general is the maturation structured model of Rotenberg [205]. Then, the physiological structure comes from a maturation velocity $\mu \in [0, 1]$, and the observable state x is the biological age, more relevant than the physical age, in other words, the degree of maturity (and then x/μ is the physical age in the previous models). Then, the density of population $n(t, x, \mu)$ satisfies the transport equation

$$\frac{\partial}{\partial t} n(t, x, \mu) + \mu \frac{\partial}{\partial x} n(t, x, \mu) + D(x, \mu) n(t, x, \mu) = \int K(x, \mu, \mu') n(t, x, \mu') d\mu',$$
(3.40)

with again boundary conditions at $x = 0$, and initial data at $t = 0$,

$$n(t, x = 0, \mu) = \int b(x', \mu', \mu) n(t, x', \mu') d\mu' dx',$$

$$n(t = 0, x, \mu) = n_0(x, \mu).$$

This model enhances stochasticity in time evolution of the population thanks to the kernel K which allows a random change of the maturation velocity, and also, as in the previous models, in the birth process. This type of equation is called a *kinetic equation* because it arises in kinetic physics, see [53, 196]. We also refer to [92] for other models with the notion of maturity and to [177] for a Generalized Relative Entropy approach.

3.9.7 Stem cells and hematopoiesis

A model of Mackey and Rey (see [163] and further papers of the same authors) has been widely studied ([3] and the references therein). It aims at describing the production of human red blood cells in the bone marrow structuring it in the age a of cells and their maturity m. Maturity $m = 0$ represents pure stem cells and should be present in order to produce the different blood cells. Mackey and Rey propose to model it with a coupled system of two nonlinear transport equations that extend the renewal equations. For $t \geq 0$, $a \geq 0$, $m \geq 0$, they write

$$\begin{cases} \frac{\partial}{\partial t} p(t, a, m) + \frac{\partial}{\partial a} p(t, a, m) + \frac{\partial}{\partial m} [V(m) p(t, a, m)] + d_1(m) p(t, a, m) = 0, \\[2mm] p(t, a = 0, m) = b_2\big(m, N(t, m)\big) N(t, m), \\[2mm] N(t, m) = \int_0^\infty n(t, a, m) da; \end{cases}$$
(3.41)

$$\begin{cases} \frac{\partial}{\partial t}n(t,a,m) + \frac{\partial}{\partial a}n + \frac{\partial}{\partial m}[V(m)n] + [d_2(m) + b_2(m, N(t,m))]n = 0, \\ n(t,a=0,m) = 2\int b_1(a,m)p(t,a,G^{-1}(m))\,da. \end{cases} \quad (3.42)$$

Here, p and n denote the population density of proliferative cells and resting cells, and D_1, d_2 their death rate. Cell division occurs according to the term with b_1 from proliferative cells (and are born in rest phase).

This system allows us to see at least two interesting features. Firstly, to avoid boundary condition at $m = 0$ (which would suppose making explicit a 'reservoir' of stem cells with its own independent dynamics, a hypothesis without biological support), it is assumed that $V(0) = 0$ and

$$\int_0^m \frac{1}{V(m')}dm' = +\infty \quad \forall m > 0, \qquad V(m) \text{ is increasing.}$$

Secondly, cell division is assumed to push backward the cells in their maturity state, which is the aim of the term $G^{-1}(m)$ with $G \in C^1([0,\infty[)$ and

$$G : \mathbb{R}^+ \to \mathbb{R}^+, \qquad \text{increasing,} \qquad G(0) = 0, \qquad G(m) < m.$$

The nonlinear term $b_2(m, N)$ is decreasing in N.

Important to notice here is that $p(t, a, m = 0)$ does not vanish but has a dynamics inherited from (3.41) by contrast with the model for the ovulary process, still structured by maturity in Section 3.9.8.

This model exhibits interesting behaviors as non-vanishing steady states and, possibly related to known diseases, periodic solutions [2].

3.9.8   Model for ovulary process

A model involving maturating cells has also been proposed by [93, 94] in order to describe the ovulary process. Here we just present a simplified version which keeps some flavor of the model and its nonlinearity.

We denote by $n(t, a, m)$ the density of granulosa cells (it should be also parametrized by the follicle under consideration and the phases should be taken into account but we do not do it for the sake of simplicity). Again t is the current time, a is the age in the cycle as usual and μ is a marker for the cell maturity. Then, the density evolution is driven by the equation

$$\begin{cases} \frac{\partial}{\partial t}n(t,a,m) + \frac{\partial}{\partial a}[g(u)n(t,a,m)] + \frac{\partial}{\partial m}[h(m,u)n(t,a,m)] + dn(t,a,m) = 0, \\ n(t,a=0,m) = 2n(t,a=1,m). \end{cases}$$
$$(3.43)$$

As in the model of hematopoiesis (Section 3.9.7), there is no boundary condition in the maturity variable because it vanishes at $m = 0$. Typically it has the form

$$h(m, u) = m(\beta(u) - m),$$

where $u = u(t) = \mathcal{U}(U(t), M(t))$ is a 'control' parameter driven by the level $U(t)$ of Follicular Stimulating Hormone (FSH in short). More precisely

$$M(t) = \int_{a=0}^{\infty} \int_{m=0}^{\infty} m \, n(t, a, m) da \, dm,$$

and

$$\frac{d}{dt} U(t) = S(M(t)) - kU(t).$$

where $S(M) = S_0 + \frac{1}{1+M(t)}$ denotes the FSH release (from pituitary gland for instance) and k its degradation rate.

3.9.9 Exercises

Exercise 1. Consider the age structured model for $B \in C^1(\mathbb{R}^+)$,

$$\begin{cases} \frac{\partial}{\partial t} n(t, x) + \frac{\partial}{\partial x} n(t, x) = 0, & t \geq 0, \ x \geq 0, \\\\ n(t, x = 0) = \int_0^{\infty} B(x) n(t, x) dx, & (3.44) \\\\ n(t = 0, x) = n^0(x) \in C_{\text{comp}}^1. \end{cases}$$

1. Give the solution $n(t, x)$ derived by the method of characteristics for the two cases $x < t$, $x > t$.

2. Assuming $n^0(0) = \int_0^{\infty} B(x) n^0(x) dx$, show that it is a C^1 solution.

3. Show that it can be reduced to the Lotka equation on $\beta(t) = n(t, x = 0)$,

$$\beta(t) = \beta^0(t) + \int_0^t B(s) \beta(t - s) ds,$$

and identify β^0.

4. Give a definition of solutions in a distribution sense for the case when the compatibility condition on $n^0(0)$ is not satisfied.

Exercise 2. For a function $V(x)$ satisfying the Cauchy–Lipschitz conditions of Section 6.1.1, consider the maturity structured model

$$\begin{cases} \frac{\partial}{\partial t} n(t, x) + \frac{\partial}{\partial x} [V(x) n(t, x)] = 0, & t \geq 0, \ x \geq 0, \\ n(t = 0, x) = n^0(x). \end{cases}$$

1. Show that it admits a solution, without need of a boundary condition at $x = 0$, if and only if

$$\int_0^{\cdot} \frac{1}{V(y)} dy = \infty.$$

2. Prove that the solution is continuous at $x = 0$ if n^0 is.

3.   a) For $V(x) = x$, prove there are eigenelements (with infinite mass), i.e., $\lambda_0 > 0$, $N > 0$ such that

$$\frac{\partial}{\partial x}\left[V(x)N(x)\right] + \lambda_0 N(x) = 0 \qquad x \geq 0;$$

   b) compute the adjoint state

$$-V\frac{\partial}{\partial x}\phi(x) + \lambda_0\phi(x) = 0 \qquad x \geq 0;$$

   c) is it possible to normalize them with $\int_0^\infty N(x)\phi(x)\,dx = 1$?

Exercise 3 (Cell division cycle with proliferative/quiescent states). The following model is based on the same type of idea as in Section 3.9.7 for cells with two states.

   Consider the model for $t \geq 0$, $x \geq 0$,

$$\begin{cases} \frac{\partial}{\partial t}p(t,x) + \frac{\partial}{\partial x}p(t,x) + B\ p(t,x) = \nu q(t,x), \\[2mm] \frac{\partial}{\partial t}q(t,x) + \frac{\partial}{\partial x}q(t,x) + \nu\ q(t,x) = 0, \\[2mm] p(t,x=0) = 0, \qquad q(t,x=0) = 2B\int_0^\infty p(t,x)dx. \end{cases} \qquad (3.45)$$

1. Give a condition on the parameters ($B > 0, \nu > 0$), $B \neq \nu$, so that there exist first eigenelements ($\lambda_0 > 0, P > 0, Q > 0, \phi_p \geq 0, \phi_q \geq 0$) and compute them explicitly, i.e., solutions to

$$\begin{cases} \frac{\partial}{\partial x}P(x) + [\lambda_0 + B]\ P(x) = \nu Q(x), \qquad \frac{\partial}{\partial x}Q(x) + [\lambda_0 + \nu]\ Q(x) = 0, \\[2mm] P(x=0) = 0, \qquad Q(x=0) = 2B\int_0^\infty P(x)dx; \end{cases}$$

$$\begin{cases} -\frac{\partial}{\partial x}\phi(x) + [\lambda_0 + B]\ \phi(x) = 2B\psi(0), \\[2mm] -\frac{\partial}{\partial x}\psi(x) + [\lambda_0 + \nu]\ \psi(x) = \nu\phi(x). \end{cases}$$

2. Give the conserved quantity and the family of generalized relative entropies for this system.

3. Give the natural weight for the L^1 estimate; Give the L^∞ estimate.

Exercise 4 (Cell division cycle with several phases). Compute the eigenelements and prove the result of Theorem 3.8. Write the Generalized Relative Entropy Inequality.

Chapter 4

Population balance equations: size structure

For unicellular organisms the renewal equation does not apply, mainly because age is not the most relevant parameter that determines mitosis (the reproduction stage). The mass of the cell, its length, its DNA content, the level of certain proteins as cyclins or some other biological parameters are often more relevant (see Figure 4.1 for a comparison between budding yeast and bacterium E. Coli).

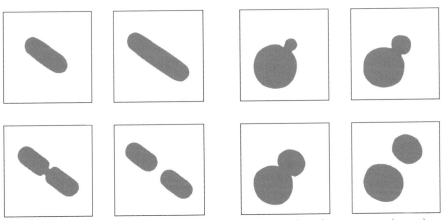

Figure 4.1: PRINCIPLE OF CELL DIVISION IN E. COLI (LEFT), AND YEAST (RIGHT).

This chapter deals with size structured models as motivated in [171], which lead to a different type of equations than age structured. It turns out that these models are much more interesting in terms of their mathematical analysis, as long as we stay away from the usual assumption 'the smallest mother cell is twice larger than the biggest daughter cell'.

We call 'equal mitosis' the case when a cell of size x divides in two cells of size $x/2$. We treat it first because the model is simpler than the general case of asymmetric division. In both cases we base our analysis on the generalized relative entropy method, once the model has been renormalized by the exponential rate inherent to its unrestricted growth. Throughout this chapter, we follow arguments taken from [198, 176, 173].

4.1   Equal mitosis

In this section we restrict ourselves to 'equal mitosis' where cells of size x divide in two equal pieces of size $x/2$ with a rate $B(x)$. Hence, we now consider the evolution equation

$$
\left\{
\begin{array}{l}
\frac{\partial}{\partial t}n(t,x) + \frac{\partial}{\partial x}n(t,x) + B(x)\,n(t,x) = 4B(2x)\,n(t,2x), \quad t > 0,\ x \geq 0, \\[2mm]
n(t, x = 0) = 0, \quad t > 0, \\[2mm]
n(0,x) = n^0(x) \geq 0.
\end{array}
\right.
$$
(4.1)

Again $B(x)$ denotes the division (birth) rate. The boundary condition here means that cells are not introduced in the system with the smallest possible size $x = 0$.

In order to explain the factor 4 in the right-hand side we have to argue that $n(t,x)dx$ is the cell density and when mitosis occurs at size $2x$ it thus creates $2 * n(2x)d(2x)$ cells. More rigorously one may consider two quantities, the total population number and its total size (mass). As it is clear from the modeling, the transport term $\frac{\partial}{\partial x}n(t,x)$ leaves the population number unchanged while the fragmentation term $4B(2x)\,n(t,2x) - B(x)\,n(t,x)$ increases it with rate $B(x)$. To see this, we integrate in x the equation and obtain

$$
\begin{aligned}
\frac{d}{dt}\int n(t,x)dx &= 4\int B(2x)\,n(t,2x)dx - \int B(x)\,n(t,x)dx \\
&= \int B(x)\,n(t,x)dx.
\end{aligned}
$$
(4.2)

Then, we consider the total mass which is increased by the transport term and left unchanged by the fragmentation term, as we see it also by integration in x after multiplication by x,

$$
\begin{aligned}
\frac{d}{dt}\int xn(t,x)dx &= \int n(t,x)dx + 4\int xB(2x)\,n(t,2x)dx - \int xB(x)\,n(t,x)dx \\
&= \int n(t,x)dx.
\end{aligned}
$$
(4.3)

In order to understand the global effect of these two partial conservation laws for each term, we consider again the eigenelements associated to this equation. They are the solution $(\lambda_0, N(x), \phi)$ to the stationary equations (again, λ_0 is sometimes called the Malthus parameter)

$$\begin{cases} \frac{\partial}{\partial x}N(x) + (\lambda_0 + B(x)) \, N(x) = 4B(2x) \, N(2x), & x \geq 0, \\ \\ N(0) = 0, \qquad N(x) > 0 \quad \text{for } x > 0, \qquad \int_0^\infty N(x)dx = 1, \end{cases} \tag{4.4}$$

$$\begin{cases} \frac{\partial}{\partial x}\phi(x) - (\lambda_0 + B(x)) \, \phi(x) = -2B(x) \, \phi(\frac{x}{2}), & x \geq 0, \\ \\ \phi(x) > 0 \quad \text{for } x \geq 0, \qquad \int_0^\infty N(x)\phi(x)dx = 1. \end{cases} \tag{4.5}$$

In this section we do not give the existence theory of solutions to equation (4.1) because it is recalled in Section 4.2.3 for the general cell division equation. Let us just point out that it provides us with weak solutions (in L^1 spaces), with the weight x, but regularity can also be proved easily for smooth B.

4.1.1 An example: B constant

By contrast with the renewal equation studied in Section 3.1, the equal mitosis equation has rarely explicit solutions and the existence theory for the eigenvalue problem (4.4)–(4.5) is more elaborate. Hence we begin with the example (see Section 4.1.2 for another) of B constant for which a solution is known explicitly (see [13]).

Lemma 4.1. *For $B(x) \equiv B$ a constant, then, the solution $(\lambda_0, N(x), \phi)$ to (4.4)–(4.5) is given by*

$$\lambda_0 = B, \qquad \phi(x) \equiv 1,$$

$$N(x) = \bar{N} \sum_{n=0}^\infty (-1)^n \, \alpha_n \, e^{-2^{n+1}Bx}, \tag{4.6}$$

with $\alpha_0 = 1$, $\alpha_n = \frac{2}{2^n - 1} \, \alpha_{n-1}$ and $\bar{N} > 0$ an appropriate normalization constant.

One can easily notice that this function N vanishes with all its derivatives at 0 and at infinity. It is depicted in Figure 4.2.

Exercise. For the fragmentation equation, with $k \in \mathbb{N}^*$,

$$\begin{cases} \frac{\partial}{\partial x}N(x) + kB \, N(x) = k^2 B \, N(kx), & x \geq 0, \\ \\ N(0) = 0, \end{cases}$$

determine the solution with a formula analogous to (4.6).

Proof. The statement on (λ_0, ϕ) is obvious and we only consider the construction on N being given that $\lambda_0 = B$.

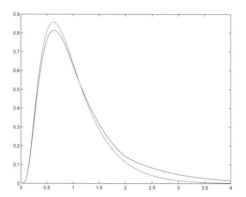

Figure 4.2: Stable size distribution given by (4.4) with B constant. The second curve shows that predation on small sizes, modeled by an additional death term $d(x)N(x)$, increases the average size.

We firstly prove that the equation is satisfied. We have

$$\frac{\partial}{\partial x}N(x) = \bar{N}\sum_{n=0}^{\infty}(-1)^n\alpha_n 2^{n+1}Be^{-2^{n+1}Bx}$$

$$= -2BN + \bar{N}\sum_{n=0}^{\infty}(-1)^n\alpha_n 2(2^n-1)Be^{-2^{n+1}Bx}$$

$$= -2BN + 2\bar{N}\sum_{n=1}^{\infty}(-1)^n 2\alpha_{n-1}Be^{-2^n B2x}$$

$$= -2BN + 4BN(2x).$$

Secondly, we prove that $N(0) = 0$. We have

$$\alpha_n = \frac{2^n}{(2^n-1)\ldots(2^1-1)}$$

so that

$$\alpha_0 - \alpha_1 = 1 - \frac{2}{2-1} = \frac{-1}{2-1},$$

and

$$\alpha_0 - \alpha_1 + \alpha_2 = \frac{-1}{2-1} + \frac{2^2}{(2^2-1)(2^1-1)} = \frac{-1}{(2^2-1)(2^1-1)}.$$

One can readily check by induction that

$$\sum_{n=0}^{k}(-1)^n\,\alpha_n = \frac{(-1)^k}{(2^k-1)\ldots(2^1-1)},$$

which proves the result as $k \to \infty$.

Thirdly, we prove the positivity of $N(x)$. Multiplying (4.4) by $\text{sgn}(N(x))$ we obtain

$$\frac{\partial}{\partial x}|N(x)| + 2B|N(x)| = 4BN(2x)\,\text{sgn}(N(x)).$$

After integration over the half line $x \geq 0$, we find

$$2B\int_0^\infty |N(x)|dx = 4B\int_0^\infty N(2x)\text{sgn}\big(N(x)\big)dx.$$

Thus, dividing by $2B$ and changing variable $y = 2x$ in the second integral, we obtain

$$\int_0^\infty |N(x)|dx = \int_0^\infty N(y)\text{sgn}\big(N(\tfrac{y}{2})\big)dy.$$

This proves that $\text{sgn}(N(x)) = \text{sgn}\big(N(\tfrac{x}{2})\big)$ for all $x > 0$.

On the other hand, the series (4.6) defining N is alternate for, say, $2Bx \geq 1$, and thus $N(x) > 0$ for large x, combined with the above sign property we conclude the positivity of N. $\qquad\square$

Uniqueness for the positive solution to (4.4) can be proved also and we refer to the more general case of Section 4.2.

The principle behind this construction can be understood in a more general framework that allows us to cover a large class of functions $B(x)$ (see [173]).

4.1.2 A counterexample: size condition on B

We consider another example which can be computed explicitly and which aims to prove that a size condition on B is still necessary as in (3.2) or (3.22). As a consequence of this section, notice however that, from this example, the size condition is no longer $\int B > 1$! A general result in this direction is stated in Section 4.2.3 (Lemma 4.4).

We choose

$$B(x) = \begin{cases} 0 & \text{for } 0 \leq x \leq 1, \\ b & \text{for } 1 \leq x \leq 2, \\ 0 & \text{for } 2 \leq x, \end{cases} \qquad (4.7)$$

and we have

Lemma 4.2. *For $b > \ln 2$ in (4.7) there is a unique solution $(\lambda_0, N(x), \phi)$ to (4.4)– (4.5) and $\lambda_0 = 0$ for $b = \ln 2$.*

For $b < \ln 2$ there is no solution.

Notice that in the case $b < \ln 2$, the solution corresponds to $\lambda_0 < 0$ and N with an exponential growth at infinity that we discard here because we cannot have $\int N = 1$.

We only consider the existence of (λ_0, N) and leave the proof as an exercise.

Exercise. Consider the value $N(1)$ and try to find the parameter λ_0 by testing the possible values λ of λ_0 in (4.4).

1. Show that a solution to (4.4) (with λ in place of λ_0) is given by

$$
N(x) = \begin{cases}
N(1)e^{\lambda+b}e^{-(\lambda+b)x} & \text{for } 1 \le x \le 2, \\
e^{-\lambda x}2b\int_1^{2x} N(y)dy & \text{for } 1/2 \le x \le 1, \\
0 & \text{for } x \le 1/2,
\end{cases}
$$

(the value for $x > 2$ does not play a role here) and conclude that a solution to (4.4) exists only if there is a solution $\lambda_0 > 0$ to

$$
\Psi(\lambda) := \lambda + 2b - 4b[e^{\lambda/2} - e^{-\lambda+b}] = 0.
$$

2. Show that for $e^b > 2$ then $\Psi(0) < 0$ and that $\Psi'(\lambda) > 1$ and conclude that there is a unique λ_0 (and N).

3. For $e^b < 2$, we write

$$
\Psi'(\lambda) := 1 + 2bX[1 - r(b)X], \qquad 0 < X = e^{-\lambda/2} < 1, \quad r(b) = 2e^{-b} > 1.
$$

Show that $\Psi(0) > 0$, and that the equation $\Psi'(\lambda) = 0$ has at most one solution λ_m and that, if this is the case, $\Psi(\lambda_m) > 1 + \lambda_m > 0$ and conclude that $\Psi(\lambda) > 0 \; \forall \lambda > 0$.

4.1.3   The conservation law and generalized relative entropy

The equation (4.1) admits also a family of entropy inequalities which generalizes the usual conservation law

$$
\frac{d}{dt}\int \phi(x)\tilde{n}(t, x)dx = 0, \tag{4.8}
$$

where we use again the notation $\tilde{n} = e^{-\lambda_0 t}n$, a function that solves

$$
\begin{cases}
\frac{\partial}{\partial t}\tilde{n}(t, x) + \frac{\partial}{\partial x}\tilde{n}(t, x) + (\lambda_0 + B(x))\,\tilde{n}(t, x) = 4B(2x)\,\tilde{n}(t, 2x), & t > 0, \; x \ge 0, \\
\tilde{n}(t, x = 0) = 0, \quad t > 0, \\
\tilde{n}(0, x) = n^0(x) \ge 0.
\end{cases}
\tag{4.9}
$$

Theorem 4.1. *Assume there exist eigenelements* (λ_0, N, ϕ) *solutions to* (4.4)–(4.5), *then for all convex and Lipschitz functions* $H : \mathbb{R} \to \mathbb{R}$ *with* $H(0) = 0$, *we have,*

$$\frac{d}{dt} \int H\left(\frac{\widetilde{n}(t,x)}{N(x)}\right) N(x)\, \phi(x) dx \leq -D_H(t) \leq 0, \qquad \forall t > 0,$$

where the entropy dissipation $D_H(t) \geq 0$ *is given by*

$$D_H(t) = -4 \int N(2x) B(2x) \phi(x) \Big[H'\left(\frac{\widetilde{n}(t,x)}{N(x)}\right)\left(\frac{\widetilde{n}(t,2x)}{N(2x)} - \frac{\widetilde{n}(t,x)}{N(x)}\right)$$
$$+ H\left(\frac{\widetilde{n}(t,x)}{N(x)}\right) - H\left(\frac{\widetilde{n}(t,2x)}{N(2x)}\right)\Big] dx \geq 0.$$

Proof. Using (4.9), we have

$$\frac{\partial}{\partial t}\frac{\widetilde{n}(t,x)}{N(x)} + \frac{\partial}{\partial x}\frac{\widetilde{n}(t,x)}{N(x)} = 4B(2x)\frac{N(2x)}{N(x)}\Big[\frac{\widetilde{n}(t,2x)}{N(2x)} - \frac{\widetilde{n}(t,x)}{N(x)}\Big],$$
$$\frac{\partial}{\partial t} H\left(\frac{\widetilde{n}(t,x)}{N(x)}\right) + \frac{\partial}{\partial x} H\left(\frac{\widetilde{n}(t,x)}{N(x)}\right) = 4B(2x)\frac{N(2x)}{N(x)} H'\left(\frac{\widetilde{n}(t,x)}{N(x)}\right)\Big[\frac{\widetilde{n}(t,2x)}{N(2x)} - \frac{\widetilde{n}(t,x)}{N(x)}\Big].$$

On the other hand

$$\frac{\partial}{\partial x}\big(N(x)\phi(x)\big) = 4\phi(x)B(2x)N(2x) - 2N(x)B(x)\phi(\tfrac{x}{2}).$$

Therefore

$$\frac{\partial}{\partial t}\Big[N(x)\phi(x)H\left(\tfrac{\widetilde{n}(t,x)}{N(x)}\right)\Big] + \frac{\partial}{\partial x}\Big[N(x)\phi(x)H\left(\tfrac{\widetilde{n}(t,x)}{N(x)}\right)\Big]$$
$$= 4B(2x)N(2x)\phi(x)H'\left(\tfrac{\widetilde{n}(t,x)}{N(x)}\right)\Big[\tfrac{\widetilde{n}(t,2x)}{N(2x)} - \tfrac{\widetilde{n}(t,x)}{N(x)}\Big]$$
$$+ \Big[4\phi(x)B(2x)N(2x) - 2N(x)B(x)\phi(\tfrac{x}{2})\Big]H\left(\tfrac{\widetilde{n}(t,x)}{N(x)}\right).$$

After integration in x we arrive at

$$\frac{d}{dt}\int_0^\infty N\phi H\left(\frac{\widetilde{n}(t,x)}{N(x)}\right) = \int_0^\infty 4\phi(x)B(2x)N(2x)H'\left(\frac{\widetilde{n}(t,x)}{N(x)}\right)\Big[\frac{\widetilde{n}(t,2x)}{N(2x)} - \frac{\widetilde{n}(t,x)}{N(x)}\Big]$$
$$+ \int_0^\infty 4\phi(x)B(2x)N(2x)\Big[H\left(\frac{\widetilde{n}(t,x)}{N(x)}\right) - H\left(\frac{\widetilde{n}(t,2x)}{N(2x)}\right)\Big] dx.$$

This is exactly the announced result. $\qquad\square$

4.1.4 Exponential time decay to Stable Size Distribution

We set again

$$\widetilde{n}(t,x) = n(t,x)e^{-\lambda_0 t}, \qquad h(t,x) = \widetilde{n}(t,x) - \int n^0(y)\phi(y)dy\, N(x).$$

As we will see in Section 4.2 the entropy inequality implies directly that $h(t)$ vanishes for $t \to \infty$ under very general conditions on $b(\cdot,\cdot)$. In this section, we give a direct exponential rate of convergence for the case $B(x) \equiv B$ (constant), therefore justifying that N is a Stable Steady Size Distribution. The following theorem is borrowed from [198] where one can also find an extension of the method to non-constant division rates B.

Theorem 4.2. *Assume $B(x) = B$, i.e., $\lambda_0 = B$, $\phi = 1$, then solutions to* (4.16) *satisfy*

$$\|h(t,x)\|_{L^1(\mathbb{R}+)} \le e^{-Bt}\big[\|h^0(x)\|_{L^1(\mathbb{R}+)} + 6B\|H^0\|_{L^1(\mathbb{R}+)}\big], \tag{4.10}$$

with

$$H^0(x) = \int_0^x h^0(y)\,dy \to 0 \quad as \quad x \to \infty.$$

Proof. We set

$$H(t,x) = \int_0^x h(t,y)dy.$$

These functions satisfy

$$\begin{cases} \dfrac{\partial}{\partial t}h(t,x) + \dfrac{\partial}{\partial x}h(t,x) + 2Bh(t,x) = 4Bh(t,2x), & t>0,\ x\ge0, \\[2mm] h(t,x=0)=0, \quad \displaystyle\int_0^\infty h(t,x)dx = 0, \quad \forall t>0, \end{cases} \tag{4.11}$$

and

$$\begin{cases} \dfrac{\partial}{\partial t}H(t,x) + \dfrac{\partial}{\partial x}H(t,x) + 2BH(t,x) = 2BH(t,2x), & t>0,\ x\ge0, \\[2mm] H(t,x=0)=0, \quad H(t,\infty)=0, \quad \forall t>0. \end{cases} \tag{4.12}$$

First step. We begin with a study of H. We have

$$\frac{\partial}{\partial t}[H(t,x)e^{Bt}] + \frac{\partial}{\partial x}[H(t,x)e^{Bt}] + B[H(t,x)e^{Bt}] = 2B[H(t,2x)e^{Bt}],$$

and thus

$$\frac{\partial}{\partial t}|H(t,x)e^{Bt}| + \frac{\partial}{\partial x}|H(t,x)e^{Bt}| + B|H(t,x)e^{Bt}| \le 2B|H(t,2x)e^{Bt}|.$$

We find after integration in x, using that H vanishes at infinity that

$$\frac{d}{dt}\int_0^\infty |H(t,x)e^{Bt}|dx \le 0, \qquad \int_0^\infty |H(t,x)|dx \le e^{-Bt}\int_0^\infty |H^0(x)|dx. \tag{4.13}$$

Second step. We work on $K(t,x) = \frac{\partial}{\partial t}H(t,x)$. We have

$$\begin{cases} \dfrac{\partial}{\partial t}K(t,x) + \dfrac{\partial}{\partial x}K(t,x) + 2BK(t,x) = 2BK(t,2x), & t>0,\ x\ge0, \\[2mm] K(t,x=0)=0, \quad K(t,\infty)=0, \quad \forall t>0. \end{cases} \tag{4.14}$$

Therefore, as in the first step, since

$$K^0(x) = -h^0(x) - 2BH^0(x) + 2BH^0(2x),$$

we deduce that

$$
\begin{aligned}
\int_0^\infty |K(t,x)|dx &\leq e^{-Bt} \int_0^\infty |K^0(x)|dx \\
&\leq e^{-Bt} \int_0^\infty \left[|h^0(x)| + 2B|H^0(x)| + 2B|H^0(2x))|\right]dx \quad (4.15) \\
&= e^{-Bt} \int_0^\infty \left[|h^0(x)| + 3B|H^0(x)|\right]dx.
\end{aligned}
$$

Third step. We deduce the time decay of h from this time decay property of H. Indeed, we compute from (4.12)

$$h(t,x) = \frac{\partial}{\partial x}H = -\frac{\partial}{\partial t}H(t,x) - 2BH(t,x) + 2BH(t,2x),$$

and thus

$$
\begin{aligned}
\int_0^\infty |h(t,x)|dx &\leq \int_0^\infty |K(t,x)|dx + 3B \int_0^\infty |H(t,x)|dx \\
&\leq e^{-Bt} \left\{ \int_0^\infty \left[|h^0(x)| + 6B \int_0^\infty |H^0(x)|dx\right\}.
\end{aligned}
$$

From this, we directly deduce the estimate of the theorem. $\qquad\square$

4.2 Size structured model for asymmetric cell division

For asymmetric division, the equation (4.1) can be generalized as (x represents again the mass or volume of the organism)

$$
\begin{cases}
\frac{\partial}{\partial t}n(t,x) + \frac{\partial}{\partial x}n(t,x) + B(x)n(t,x) = \int_x^\infty b(x,y)n(t,y)dy, & t \geq 0, \ x \geq 0, \\[2mm]
n(t, x = 0) = 0, & t \geq 0, \\[2mm]
n(t = 0, x) = n^0(x),
\end{cases}
$$

$$(4.16)$$

which means that a mother cell of size $y \geq 0$ divides in two daughter cells of sizes $x \geq 0$ and $x - y \geq 0$ with rate $b(x,y)$.

For consistency with the modeling one has to impose (it turns out that it is easier now to invert the variables x and y, and thus to consider the division rate of a cell with size y giving two cells of size x and $y - x$)

$$b(x,y) \geq 0, \qquad b(x,y) = 0 \quad \text{for} \ y < x, \tag{4.17}$$

$$\int b(x,y)dx = 2B(y), \tag{4.18}$$

$$\int xb(x,y)dx = yB(y), \tag{4.19}$$

$$b(x, y) = b(y - x, y). \tag{4.20}$$

The first assumption takes into account that after division we should have cells of nonnegative sizes and thus $y - x \geq 0$, the second allows us to say that division occurs giving two cells since, after integration in x we have

$$\frac{d}{dt} \int n(t, x) dx = \int \int b(x, y) n(t, y) dy\, dx - \int B(x) n(t, x) dx = \int B(y) n(t, y) dy. \tag{4.21}$$

The third assumption takes into account mass conservation in the division process; we have, multiplying equation (4.16) by x, and integrating

$$\frac{d}{dt} \int x n(t, x) dx - \int n(t, x) dx + \int x B(x) n(t, x) dx = \int \int x b(x, y) n(t, y) dy\, dx.$$

and thus

$$\frac{d}{dt} \int x n(t, x) dx = \int n(t, x) dx. \tag{4.22}$$

The last assumption just expresses that the model is the same by accounting only for the size density $n(t, x)$ or $n(t, y - x)$.

This equation also arises in physics to describe fragmentation processes, [161, 30, 27]. Then the drift (growth) term $\frac{\partial}{\partial x} n(t, x)$ is not present and this is a fundamental difference. Also, there is no reason to stick with assumption (4.18) and the number of fragments could be anything (larger than 2 say).

We can recover the renewal equation and the equal mitosis equation as two particular examples of this equation, for appropriate choice of b,

$$b(x, y) = B(y)\, [\delta(x = y) + \delta(x = 0)], \qquad \text{(renewal equation)}, \tag{4.23}$$

$$b(x, y) = 2B(y)\, \delta(x = y/2), \qquad \text{(equal mitosis)}. \tag{4.24}$$

These choices satisfy the assumptions (4.17)–(4.20). More generally, one can consider, for a parameter $0 \leq \sigma \leq 1$, the case

$$b(x, y) = B(y)\, [\delta(x = \sigma y) + \delta(x = (1 - \sigma)y)], \qquad \text{(general mitosis)}. \tag{4.25}$$

4.2.1   Existence, regularity and comparison principle

A general, and suboptimal, result can be proved in applying the Banach-Picard fixed point method as we did in Section 3.3. We state it for the sake of completeness but do not (re)prove it (see Sections 3.3, 6.6.2).

Theorem 4.3. *Assume that $B \in L^\infty(\mathbb{R}^+)$, $B \geq 0$ and that $b(\cdot, \cdot)$ is a measure satisfying (4.17)–(4.18), and $(1 + |x|) n^0(x) \in L^1(\mathbb{R}^+)$, then there is a unique solution in distribution sense $n \in C(\mathbb{R}^+; L^1(\mathbb{R}^+))$ to (4.16) and we have*

$$n_1^0 \leq n_2^0 \quad \Longrightarrow \quad n_1(t, x) \leq n_2(t, x),$$

$$\|n(t)\|_{L^1(\mathbb{R}^+)} \leq \|n^0\|_{L^1(\mathbb{R}^+)} e^{\|B\|_{L^\infty} t},$$

$$\frac{d}{dt} \int n(t,x)dx = \int B(x)n(t,x)dx.$$

If (4.19) also holds, then, with $m^0 = \int n^0(x)dx$,

$$\int xn(t,x)dx = \int xn^0(x)dx + t\, m^0 \qquad \forall t \geq 0.$$

The natural spaces for existence and contraction principle are however somewhat different. Using some stronger assumptions, we also have

Theorem 4.4. *With the same assumptions as in Theorem 4.3 and those of Theorem 4.6 for the existence of eigenelements (λ_0, N, ϕ), there is a unique solution $n \in C(\mathbb{R}^+; L^1(\phi(x)dx))$ to (4.16) and we have*

$$\int_{\mathbb{R}^+} n(t,x)e^{-\lambda_0 t}\phi(x)dx = \int_{\mathbb{R}^+} n^0(x)\phi(x)dx,$$

$$\int_{\mathbb{R}^+} |n(t,x)|e^{-\lambda_0 t}\phi(x)dx = \int_{\mathbb{R}^+} |n^0(x)|\phi(x)dx,$$

$$C_- N(x) \leq n^0(x) \leq C_+ N(x) \implies C_- N(x) \leq n(t,x)e^{-\lambda_0 t} \leq C_+ N(x).$$

These results follow again from the Generalized Relative Entropy principle which we give below for the cell division equations. We refer to Sections 6.3, 6.4 for a general derivation of this type of results.

Also for later purpose, we can mention a regularity result similar to that in Theorem 3.2 for age structure

Theorem 4.5. *With the same assumptions as in Theorem 4.4 and with initial data satisfying*

$$|n^0(x)| \leq C_0 N(x), \qquad \frac{\partial}{\partial x}n^0(x) \in L^1(\phi(x)dx),$$

the solution to (4.16) satisfies, setting $\tilde{n} = ne^{-\lambda_0 t}$,

$$\int_0^\infty |\frac{\partial}{\partial t}\tilde{n}(t,x)|\phi(x)dx \leq C(n^0) \qquad \forall t \geq 0, \qquad (4.26)$$

$$\int_0^\infty |\frac{\partial}{\partial x}\tilde{n}(t,x)|\phi(x)dx \leq C_1(n^0) \qquad \forall t \geq 0. \qquad (4.27)$$

It is more demanding to establish L^∞ bounds for those derivatives; this requires a control by $CN(x)$ of the initial birth term $\int_x^\infty b(x,y)n^0(y)dy$, a condition that is incompatible with the examples 4.23–4.23 we have in mind.

As it is, Theorem 4.5 provides Bounded Variation (BV in short) regularity of the solution. This low regularity is compatible with discontinuities and thus with the discontinuity at $x = 0$ in the age structured case. It also provides local strong compactness of families of solutions, see for instance [41, 71, 99].

Proof. First step. Time derivative. We recall the equation

$$\frac{\partial}{\partial t}\widetilde{n}(t,x) + \frac{\partial}{\partial x}\widetilde{n}(t,x) + (\lambda_0 + B(x))\widetilde{n}(t,x) = \int_x^\infty b(x,y)\widetilde{n}(t,y)dy.$$

We obtain the equation on $q(t,x) = \frac{\partial}{\partial t}\widetilde{n}(t,x)$, by differentiating it in time, and thus q satisfies the same equation. From the contraction principle in Theorem 4.4, we conclude

$$\int_0^\infty |q(t,x)|\phi(x)dx \le \int_0^\infty |q(t=0,x)|\phi(x)dx.$$

But

$$q(t=0,x) = -\frac{\partial}{\partial x}n^0(x) - (\lambda_0 + B(x))n^0(x) + \int_x^\infty b(x,y)n^0(y)dy.$$

We may bound $|n^0|$ by $C_0 N$, replace $\int_x^\infty b(x,y)N(y)dy$ by the other terms of the equation on N and we arrive at

$$\int_0^\infty |q(t=0,x)|\phi(x)dx \le \int_0^\infty \left[|\frac{\partial}{\partial x}n^0(x)| + |\frac{\partial}{\partial x}N(x)|\right]\phi(x)dx + 2C_0(\lambda_0 + B_M).$$

Second step. Space derivative. We have

$$\frac{\partial}{\partial x}\widetilde{n}(t,x) = -\frac{\partial}{\partial t}\widetilde{n}(t,x) - (\lambda_0 + B(x))\widetilde{n}(t,x) + \int_x^\infty b(x,y)\widetilde{n}(t,y)dy.$$

The control of $\frac{\partial}{\partial t}\widetilde{n}(t,x)$ in the first step and $|\widetilde{n}(t,x)|\|eq C_0 N$ (see Theorem 4.4) gives us a control similar to that on the time derivative. $\square$

4.2.2 Generalized relative entropy (1)

We follow our computation of Section 6.4 and begin with a general abstract calculation that does not involve the eigenelements. Namely we consider a solution $\psi(t,x)$ to the dual equation

$$\frac{\partial}{\partial t}\psi(t,x) + \frac{\partial}{\partial x}\psi(t,x) - B(x)\psi(t,x) = -\int_0^\infty b(y,x)\psi(t,y)dy, \qquad t \ge 0, \ x \ge 0.$$

As in the parabolic case, this equation should be understood as a backward problem. Also we do not assume any relation between B and $b(x,y)$ in this section.

Lemma 4.3. *For smooth functions n, p, ψ, with sufficient decay in x at infinity and for $n(t,x)$ (no specific sign), $p(t,x) > 0$ solutions to (4.16) and $\psi(t,x) \ge 0$ as above, we have, for all convex functions $H : \mathbb{R} \to \mathbb{R}$,*

$$\frac{d}{dt}\int \psi(t,x)p(t,x)H\left(\frac{n(t,x)}{p(t,x)}\right)dx = -D^H(t) \le 0,$$

$$D^H(t) = \int\int \psi(t,x)p(t,y)b(x,y)\{H\left(\frac{n(t,y)}{p(t,y)}\right) - H\left(\frac{n(t,x)}{p(t,x)}\right) \qquad (4.28)$$

$$- H'\left(\frac{n(t,x)}{p(t,x)}\right)\left[\frac{n(t,y)}{p(t,y)} - \frac{n(t,x)}{p(t,x)}\right]\}dx\,dy.$$

We have in mind two possible cases of interest for the choice of p and ψ. The first one is the periodic case along with Floquet theory (see [176] and Section 6.3.2). The other special case here is to take $p(t,x) = Ne^{\lambda_0 t}$ and $\psi = \phi e^{\lambda_0 t}$ where N and ϕ are the eigenelements associated with the eigenvalue λ_0 (see Section 4.2.3). Then we arrive at

$$\frac{d}{dt}\int \phi(x)N(x)H\left(\frac{n(t,x)e^{-\lambda_0 t}}{N(x)}\right)dx = -D^H(t) \le 0,$$

$$D^H(t) = \int\int \phi(x)N(y)b(x,y)\left\{ H\left(\frac{n(t,y)e^{\lambda_0 t}}{N(y)}\right) - H\left(\frac{n(t,x)e^{\lambda_0 t}}{N(x)}\right) \right.$$
$$\left. -H'\left(\frac{n(t,x)e^{\lambda_0 t}}{N(x)}\right)\left[\frac{n(t,y)e^{\lambda_0 t}}{N(y)} - \frac{n(t,x)e^{\lambda_0 t}}{N(x)}\right]\right\}dx\,dy.$$
$$(4.29)$$

This is the result that is used below to explain the long time behavior of solutions to the cell division equation.

Proof. We now skip the time dependency in our notation and compute successively (leaving the details to the reader)

$$\frac{\partial}{\partial t}\left(\frac{n(x)}{p(x)}\right) + \frac{\partial}{\partial x}\left(\frac{n(x)}{p(x)}\right) = \int b(x,y)\frac{p(y)}{p(x)}\left[\frac{n(y)}{p(y)} - \frac{n(x)}{p(x)}\right]dy, \qquad (4.30)$$

and thus

$$\frac{\partial}{\partial t}H\left(\frac{n(x)}{p(x)}\right) + \frac{\partial}{\partial x}H\left(\frac{n(x)}{p(x)}\right) = H'\left(\frac{n(x)}{p(x)}\right)\int b(x,y)\left[\frac{n(y)}{p(x)} - \frac{p(y)n(x)}{p^2(x)}\right]dy.$$

But we also have

$$\frac{\partial}{\partial t}\left(\psi(x)p(x)\right) + \frac{\partial}{\partial x}\left(\psi(x)p(x)\right) = \psi(x)\int b(x,y)p(y)dy - p(x)\int b(y,x)\psi(y)dy,$$

and thus

$$\frac{\partial}{\partial t}\left(\psi(x)p(x)H\left(\frac{n(x)}{p(x)}\right)\right) + \frac{\partial}{\partial x}\left(\psi(x)p(x)H\left(\frac{n(x)}{p(x)}\right)\right)$$
$$= \psi(x)p(x)H'\left(\frac{n(x)}{p(x)}\right)\int b(x,y)\left[\frac{n(y)}{p(x)} - \frac{p(y)n(x)}{p^2(x)}\right]dy$$
$$+ H\left(\frac{n(x)}{p(x)}\right)\int \left[b(x,y)\psi(x)p(y) - b(y,x)\psi(y)p(x)\right]dy.$$

Integrating in x this identity and inverting the variables x and y in the very last term, we obtain the announced equality. $\qquad \square$

4.2.3 Eigenelements

As we did for the renewal equation and for equal mitosis, we now consider the first eigenelements (and in particular the Malthus parameter λ_0)

$$\begin{cases} \frac{\partial}{\partial x}N(x) + (\lambda_0 + B(x))N(x) = \int_x^\infty b(x,y)N(y)dy, & x \ge 0, \\ N(x=0) = 0, \qquad N(x) > 0 \text{ for } x > 0, \qquad \int N = 1; \end{cases} \qquad (4.31)$$

$$\begin{cases} \frac{\partial}{\partial x}\phi(x) - (\lambda_0 + B(x))\phi(x) = -\int_0^\infty b(y,x)\phi(y)dy, & x \geq 0, \\ \phi(x) > 0, \qquad \int \phi N = 1. \end{cases} \tag{4.32}$$

As pointed out by the examples in the renewal and equal mitosis cases, a (large) size condition on B is needed for existence of these eigenelements. A general argument can convince us of this and gives

Lemma 4.4. *If a solution to (4.4) exists, then*

$$\int_0^\infty B(x)dx \geq 1/2. \tag{4.33}$$

Proof. Firstly, after integration in x we obtain (using (4.17))

$$\lambda_0 = \int B(x)N(x)dx \geq 0.$$

Secondly, integrating again, but between 0 and x, we find

$$\begin{aligned} N(x) &\leq \int_{z=0}^x \int_{y=x}^\infty b(z,y)N(y)dydz \leq \int_{y=0}^\infty \int_{z=0}^y b(z,y)N(y)dydz \\ &= 2\int B(y)N(y). \end{aligned} \tag{4.34}$$

Therefore

$$\|N(x)\|_{L^\infty(\mathbb{R}^+)} \leq 2\int B(x)dx \, \|N\|_{L^\infty(\mathbb{R}^+)},$$

and if there is a solution, then we should have

$$\int B(x)dx \geq 1/2. \qquad \square$$

Our assumptions below are stronger since they imply that

$$\int B = \infty, \qquad \|xB(x)\|_{L^\infty(\mathbb{R}^+)} = \infty.$$

Another heuristic reason for the need of a large enough division rate B is that for $B \equiv 0$, there is no solution.

Theorem 4.6. *Assume that $b(\cdot,\cdot)$ is a measure satisfying (4.17)–(4.18), and that $B \in L^\infty(\mathbb{R}^+)$, satisfies, for some constants $0 < B_m \leq B_M$,*

$$\begin{cases} \forall x \geq 0, \quad B(x) \leq B_M < \infty, \\ 0 < B_m \leq B(x), \quad \forall x > x_-, \end{cases} \tag{4.35}$$

then there is a unique solution in distribution sense (λ_0, N, ϕ) to (4.31), (4.32) and we have

$$0 < \lambda_- := b_m(1 + 2b_m x_-) \leq \lambda_0 \leq B_M,$$

$$\forall \mu < \lambda_0, \qquad \int e^{\mu x} N(x) dx \le \frac{\lambda_0}{\lambda_0 - \mu}, \qquad \text{and } e^{\mu x} N(x) \in L^\infty(\mathbb{R}^+),$$

$$\frac{\partial}{\partial x} e^{\mu x} N(x) \in L^1(\mathbb{R}^+), \qquad \forall \mu < \lambda_0,$$

$$\frac{\phi(x)}{1 + x} \in L^\infty(\mathbb{R}^+), \qquad \frac{\partial}{\partial x} \phi \in L^\infty_{\text{loc}}(\mathbb{R}^+).$$

These results can be improved (see [173]) in terms of assumptions and estimates, in particular one can find there the case where B has compact support. Variants are also given in Section 4.2.5. There, one can also deal with B that decays to 0 for x large. an assumption which is reasonable because one could wish to include the fact that for large x cells do not divide.

As noticed in Section 4.1.1, for $B(x) \equiv B$ (a constant) then $\lambda_0 = B$ and $\phi \equiv 1$, because thanks to assumption (4.18) we have indeed

$$\int_0^\infty b(x, y) \, 1 \, dy = 2B.$$

But the equation on N never has explicit solutions.

Proof. We use the truncated equation presented in Section 6.6.2 and which can be solved by the Krein-Rutman theorem (see Theorem 6.6). We prove uniform estimates on this model (6.39) having in mind that we choose truncation parameters $\varepsilon \to 0$, $R = R_\varepsilon \to \infty$, $b_\varepsilon(x, y) \to b(x, y)$ (in the sense of weak convergence to measures), and $B_\varepsilon(y) = \frac{1}{2} \int_0^y b_\varepsilon(x, y) dx$. We would like to point out four difficulties to keep in mind;

(i) A specific difficulty related to the approximation with b smooth that implies $B_\varepsilon(0) = 0$, a constraint that might disappear in the limit.

(ii) It is obvious that these estimates allow to pass to the limit, strongly in all $L^p(\mathbb{R}^+)$, in the solutions N_ε and ϕ_ε. Notice that even though b is a measure in the limit, solutions are well defined (and stable) in the weak sense because after testing $b(x, y)$ we obtain quantities as $\int b(x, y) q(x) dx$ which are bounded functions in y.

(iii) The worst case of b is the renewal equation, then we have already mentioned that N has discontinuity at $x = 0$. It also proves some kind of optimality in the above stated estimates.

(iv) By comparison with N, $\phi(x)$ can vanish as we know again from the renewal equation.

Because we are going to work on it now, we recall the truncated equation (6.39) for N, dropping the index ε in N and R,

$$\begin{cases} \frac{\partial}{\partial x} N(x) + \left(\lambda_\varepsilon + B(x) \right) N(x) - \int_0^R b(x, y) N(y) dy = 0, & 0 \le x \le R, \\ \\ N(x = 0) = \varepsilon, \qquad N(x) > 0 \qquad \int_0^R N(x) dx = 1. \end{cases} \tag{4.36}$$

First estimate. We integrate on $[0, R]$ and find

$$N(R) + \lambda_\varepsilon = N(0) + \int_0^R B(y)N(y)dy \le \varepsilon + B_M,$$

and thus

$$\lambda_\varepsilon \le \varepsilon + B_M. \tag{4.37}$$

Similarly, integrating between 0 and x, we find

$$N(x) \le N(0) + \int_0^x \int_x^z b(z,y)N(y)dydz$$

$$\le \varepsilon + 2\int_0^\infty B(y)N(y)dy = -\varepsilon + 2\lambda_\varepsilon + 2N(R). \tag{4.38}$$

Second estimate. We multiply by x and integrate on $[0, R]$ and find

$$RN(R) + \lambda_\varepsilon \int_0^R xN(x)dx + \int_0^R xB(x)N(x)dx$$
$$= \int_0^R N(x)dx + \int_0^R \int_0^R xb(x,y)N(y)dydx,$$

$$RN(R) + \lambda_\varepsilon \int_0^R xN(x)dx = 1.$$

Thus $N(R) \le 1/R$ and coming back to the first estimate we also find

$$\lambda_\varepsilon \ge \varepsilon - \tfrac{1}{R} + \int_0^R B(y)N(y)dy$$
$$\ge \varepsilon - \tfrac{1}{R} + \int_{x_-}^R b_m N(y)dy$$
$$\ge \varepsilon - \tfrac{1}{R} + b_m\big(1 - \int_0^{x_-} N(y)dy\big)$$
$$\ge \varepsilon - \tfrac{1}{R} + b_m(1 - x_-(-\varepsilon + 2\lambda_\varepsilon + 2N(R)))$$

(see (4.38))) and we arrive, for ε small enough and R large enough, at

$$(1 + 2x_- b_m)\lambda_\varepsilon \ge \varepsilon - \tfrac{1}{R} + b_m(1 - 2x_- N(R)),$$

which gives an explicit lower bound

$$\lambda_\varepsilon \ge \lambda_{\varepsilon,-} = [\varepsilon - \tfrac{1}{R} + b_m(1 - 2x_- N(R))]/[(1 + 2x_- b_m)]. \tag{4.39}$$

Third estimate. We multiply by $e^{\mu x}$ and integrate on $[0, R]$ and find

$$N(R)e^{\mu R} + (\lambda_\varepsilon - \mu)\int_0^R e^{\mu x}N(x)dx + \int_0^R e^{\mu x}B(x)N(x)dx$$
$$= \varepsilon + \int_0^\infty \int_{x=0}^y e^{\mu x}b(x,y)dxN(y)dy$$
$$= \varepsilon + \int_0^\infty \int_{x=0}^y (1 + \mu x + \cdots + \tfrac{(\mu x)^n}{n!} + \cdots)dxN(y)dy$$
$$\le \varepsilon + \int_0^\infty \int_{x=0}^y (1 + \mu x + \cdots + \tfrac{x\mu^n y^{n-1}}{n!} + \cdots)dxN(y)dy$$
$$\le \varepsilon + \int_0^\infty B(y)(2 + \mu y + \cdots + \tfrac{(\mu y)^n}{n!} + \cdots)N(y)dy$$
$$= \varepsilon + \int_0^\infty (1 + e^{\mu y})B(y)N(y)dy.$$

As a consequence we find (see also the first line of the first estimate, and $N(R) \leq 1/R$)

$$\int_0^R e^{\mu x} N(x) dx \leq (\lambda_\varepsilon + \frac{1}{R})/(\lambda_\varepsilon - \mu). \qquad (4.40)$$

Fourth estimate. Arguing as in the third step but after integration between 0 and x, we also find that

$$e^{\mu x} N(x) \leq (\lambda_\varepsilon + \tfrac{1}{R}) + b_M \ (\lambda_\varepsilon + \tfrac{1}{R})/(\lambda_\varepsilon - \mu),$$

and similarly we can multiply by $e^{\mu x}$ and find by the chain rule that $\frac{\partial}{\partial x} e^{\mu x} N(x) \in L^1(\mathbb{R}^+)$.

At this stage we have obtained the relevant estimates on N in (4.36) and we turn to the estimates on the adjoint equation

$$\begin{cases} -\frac{\partial}{\partial x}\phi(x) + (\lambda_\varepsilon + B(x))\phi(x) - \int_0^R b(y, x)\phi(y) dy = \varepsilon\phi(0), & 0 \leq x \leq R, \\ \\ \phi(x = R) = 0, \qquad \phi(x) \geq 0 \qquad \int_0^R \phi(x)N(x) dx = 1. \end{cases} \qquad (4.41)$$

Fifth estimate. From this equation we deduce that

$$\frac{\partial}{\partial x}\left(\phi(x)e^{-\int_0^x (B+\lambda_\varepsilon)}\right) \leq 0,$$

therefore, for some constant $C(A)$ independent of ε we have

$$\forall A > 0, \ \exists C(A) \quad \text{such that} \quad \phi(x) \leq C(A)\phi(0). \qquad (4.42)$$

Sixth estimate. Finally we notice that the equation (4.41) can be seen as a backward transport equation and therefore satisfies the maximum principle and we can easily build, following [173], an affine supersolution $\bar{\phi}$ that is positive at $x = R$, therefore $\phi(x) \leq \bar{\phi}(x)$.

This we cannot do on $[0, R]$, but only on a subinterval $[A_0, R]$ with the function $\bar{\varphi}(x) = x + \nu$ where

$$A_0 = 2/\lambda_\varepsilon, \qquad \nu = 1/(B_M - \lambda_\varepsilon).$$

Indeed we have to check that, on $[A_0, R]$ we have

$$-\frac{\partial}{\partial x}\bar{\varphi} + (\lambda_\varepsilon + B)\bar{\varphi} = -1 + (\lambda_\varepsilon + B)\bar{\varphi} \geq \int_0^R b(y, x)\bar{\varphi}(y) dy = B(x + 2\nu),$$

and this follows from our choice for A_0, ν.

Then we conclude the sublinearity of ϕ by choosing the supersolution $\bar{\phi}(x) = K\phi(0)\bar{\varphi}$ with K large enough so that

$$\bar{\phi}(x) \geq \phi(x) + \varepsilon/\lambda_\varepsilon \qquad \text{on } [0, A_0].$$

Therefore we have on $[A_0, R]$,

$$-\frac{\partial}{\partial x}\bar{\phi}(x) + \left(\lambda_\varepsilon + B(x)\right)\bar{\phi}(x) \geq \varepsilon\phi(0) + \int_{A_0}^R b(y,x)\bar{\phi}(y)dy + \int_0^{A_0} b(y,x)\phi(y)dy,$$

which is a supersolution to the equation (the same) satisfied by ϕ. Therefore $\phi \leq \bar{\phi}$ and the result is proved. Then the local Lipschitz regularity of ϕ follows directly from the equation (4.41). $\qquad\qquad\qquad\qquad\qquad\qquad\qquad\qquad\qquad\qquad\qquad\qquad\Box$

4.2.4   Trend to a Stable Size Distribution

We are now ready to study under which circumstances the renormalized size distribution $n(t,x)e^{-\lambda_0 t}$ can converge in large time to a distribution called the Stable Size Distribution. This is not always true because the details of the size division rate $b(x,y)$ is important here and we borrow from [176] the condition: there exists a C^1 function $\Gamma : (0,\infty) \to (0,\infty)$ such that

$$\begin{cases} \{(y = \Gamma(x), \quad x \geq 0\} \subseteq \Delta = \mathrm{Supp}_{[0,\infty[\times[0,\infty[}\ b(x,y), \\[2mm] \frac{\partial}{\partial x}\Gamma(x) \neq 1 \ \forall x \neq 0. \end{cases} \qquad (4.43)$$

This condition has to be understood taking into account that b can be a measure. For instance in the equal mitosis case, we take $\Gamma(x) = x/2$ and this non-degeneracy condition means $B > 0$. In the case of the renewal equation, $\Gamma(x) = 0$ and again non-degeneracy boils down to $B > 0$. As in this case, it is possible to improve the condition in order to include the cases where B vanishes but we prefer to keep simplicity.

Notice also that we only know exponential rates of convergence in specific cases: renewal equation as already mentioned ([100, 171, 124]) or assumption (smallest mother is bigger than the largest daughter cell), equal mitosis, see [198] and its account in Section 4.1.4. Again we recall that the main difficulty is that, because $b(x,y)$ vanishes for $y < x$, we cannot hope for a Poincaré inequality that controls some generalized relative entropy from its dissipation rate, in contrast with the case of full scattering equation (Sections 6.3 and 6.4.3).

Theorem 4.7. *We make the assumptions of Theorem 4.4 and suppose the non-degeneracy condition (4.43) on the support of b. Then, the solutions to (4.16) tend to a steady state. Namely, with $\rho = \int_0^\infty n^0(x)\phi(x)dx$,*

$$\lim_{t\to\infty}\int_0^\infty |n(t,x)e^{-\lambda_0 t} - \rho\,N(x)|\,\phi(x)\,dx = 0. \qquad (4.44)$$

Proof. First step. We set

$$h(t,x) = n(t,x)e^{-\lambda_0 t} - \rho\,N(x).$$

We first notice that $h(t, x)$ being a solution to the cell division equation (4.16), the contraction principle in Theorem 4.4 shows that

$$\int_0^\infty |n(t, x)e^{-\lambda_0 t} - \rho \, N(x)| \, \phi(x) \, dx \Downarrow L, \qquad t \to \infty.$$

And it remains to show that $L = 0$. As we mentioned for age structured models (see Section 3.6), the contraction property allows us to do so, by density, for the BV solutions given by Theorem 4.5, i.e., for those solutions that satisfy for all $t \geq 0$,

$$|h| \leq C_0 N, \quad \int_0^\infty |\frac{\partial}{\partial t} h(t, x)|\phi(x)dx \leq C(n^0) \quad \int_0^\infty |\frac{\partial}{\partial x} h(t, x)|\phi(x)dx \leq C(n^0). \tag{4.45}$$

Second step. We then introduce the sequence of functions $h_n(t, \cdot) = h(t + t_n, \cdot)$. After extracting a subsequence, still denoted h_n, we have $h_n \to g$ strongly in $L^1([0, T] \times \mathbb{R}^+)$ for all $T > 0$, because of the global BV regularity on h proved in Theorem 4.5 (see also the comments after this theorem). And we have that g is a solution to the cell division equation (4.16) and

$$|g(t, x)| \leq C_0 N(t, y).$$

Third step. We can now work on the entropy dissipation of $h(t, x)$. From the Generalized Relative Entropy inequality (4.29), we have, using the square entropy $H(u) = u^2$,

$$\int_0^\infty \int_0^\infty \phi(x)b(x, y)N(y)|\frac{h(t,x)}{N(x)} - \frac{h(t,y)}{N(y)}|^2 dxdy \, dt \leq C.$$

Therefore, as $n \to \infty$,

$$\int_0^\infty \int_0^\infty \phi(x)b(x, y)N(y)|\frac{h_n(t,x)}{N(x)} - \frac{h_n(t,y)}{N(y)}|^2 dxdy \, dt$$
$$= \int_n^\infty \int_0^\infty \phi(x)b(x, y)N(y)|\frac{h(t,x)}{N(x)} - \frac{h(t,y)}{N(y)}|^2 dxdy \, dt \to 0.$$

By the strong limit of h_n we arrive at

$$\int_0^\infty \int_0^\infty \phi(x)b(x, y)N(y)|\frac{g(t,x)}{N(x)} - \frac{g(t,y)}{N(y)}|^2 dxdy \, dt = 0. \tag{4.46}$$

In other words

$$\frac{g}{N}(t, \Gamma(x)) = \frac{g}{N}(t, x), \qquad \forall t > 0, \ x \geq 0. \tag{4.47}$$

On the other hand, in the limit the entropy dissipation for g/N vanishes in (4.30) (recall p stands for N in the case at hand) and thus the division part of (4.30) vanishes for this g/N and we obtain

$$\frac{\partial}{\partial t} \frac{g}{N} + \frac{\partial}{\partial x} \frac{g}{N} = 0. \tag{4.48}$$

Fourth step. Thanks to Lemma 4.5 below we have $g(t, x) = Cst$ and the mass condition $\int g(t, x)\phi(x)dx = 0$ allows us to conclude $g = 0$ and thus $L = 0$. □

Lemma 4.5. *Any function $u = g/N$ satisfying (4.48), (4.47) is constant.*

Proof. On the one hand, we have

$$(\partial_t u)(t, x) = (\partial_t u(t, \Gamma(x))) = (\partial_t u)(t, \Gamma(x)). \qquad (4.49)$$

On the other hand, we have

$$(\partial_x u)(t, x) = (\partial_x u(t, x)) = (\partial_x u(t, \Gamma(x))) = \Gamma'(x)(\partial_x u)(t, \Gamma(x)). \qquad (4.50)$$

We deduce gathering (4.49), (4.50) and using (4.48) that

$$(\partial_t u)(t, \Gamma(x)) + \Gamma'(x)(\partial_x)u(t, \Gamma(x)) = 0, \qquad \forall t > 0, \ x \geq 0, \qquad (4.51)$$

and from (4.48) we also have

$$(\partial_t u)(t, \Gamma(x)) + (\partial_x)u(t, \Gamma(x)) = 0, \qquad \forall t > 0, x \geq 0. \qquad (4.52)$$

Combining (4.51), (4.52) we get

$$(\Gamma'(x) - 1)(\partial_x)u(t, \Gamma(x)) = 0,$$

from which we deduce, since $\Gamma'(x) \neq 1$,

$$(\partial_x)u(t, x) = \Gamma'(x)(\partial_x)u(t, \Gamma(x)) = 0.$$

Finally using again the transport equation (4.48) we obtain indeed that u is constant. $\qquad \square$

4.2.5   Some extensions and exercises

Exercise. We can relax the assumption $0 < B_m \leq B(x)$ in Theorem 4.6. We define $\alpha(x)$ by

$$\int y^2 b(x, y) dy = (1 - \alpha(x))x^2 B(x),$$

and assume that, as $x \to \infty$, $B(x)$ decreases to 0 but not too fast, namely

$$\alpha(x) \, x^2 \, B(x) - (1 + \theta)x \geq O(1), \qquad \text{with } \theta > 0. \qquad (4.53)$$

1. For equal mitosis, show that $\alpha(x) \equiv 1/2$.

2. For a solution (λ_0, N) to (4.31) with enough decay in x at infinity, show that

$$\int BN = \lambda_0, \qquad \int x N(x) = 1/\lambda_0,$$

   and

$$\lambda_0^2 \int x^2 N(x) + \lambda_0 \int \alpha(x) x^2 B(x) N(x) = 2.$$

3. Show that $\lambda_0^2 \int x^2 N(x) \geq 1$, and $\lambda_0 \int \alpha(x) x^2 B(x) N(x) \geq 1 + \theta + O(\lambda_0)$, and prove a lower bound on λ_0.

4. Conclude that there exists a solution (λ_0, N) to (4.31).

Exercise. In the framework on the exercise above, for $0 < \beta < 1$, we define

$$\int y^{1+\beta} b(x,y) dy = (1 - \alpha_\beta(x)) x^2 B(x),$$

and assume that, as $x \to \infty$,

$$\alpha_\beta(x)\, x^{1+\beta}\, B(x) - (1 + \theta_\beta) x^\beta \geq O(1), \qquad \text{with} \;\; \theta_\beta > 0.$$

1. Show that (this is always true)

$$(1 + \beta) \int x^\beta N(x) = \lambda_0 \int x^{1+\beta} N(x) + \int x^{1+\beta} \alpha_\beta(x) B(x) N(x),$$

$$\int x^{1+\beta} \alpha_\beta(x) B(x) N(x) \leq \beta \int x^\beta N(x).$$

2. Prove, with the assumption, a lower bound on λ_0. Conclude that there exists a solution (λ_0, N) to (4.31).

3. For equal mitosis, compare the possible constants γ in the time decay rate $B(x) \leq \gamma/(1+x)$ for this method and that of the exercise above.

Exercise. Consider the equation, with $B(x)$ continuous and satisfying $0 < B_m \leq B(x) \leq B < \infty$,

$$\begin{cases} \varepsilon \frac{\partial}{\partial x} N_\varepsilon + (\lambda_e + B) N_\varepsilon = \int_x^\infty b(x,y) N_\varepsilon(y) dy, \\[2mm] N_\varepsilon(0) = 0, \qquad \int N_\varepsilon = 1, \end{cases}$$

and prove that as $\varepsilon \to 0$

$$\lambda_e \to B(0), \qquad N_\varepsilon \to \delta(x = 0).$$

Hint: Integrate in dx and $x\, dx$.

4.3 Population balance equations: other examples, nonlinear examples

Several other models arise to describe populations in various contexts. In this section we present some of them, taken from several fields of applications. We do not present any specific analysis, which can be carried out by methods similar to those we have presented before.

4.3.1   Daphnia (size structured models with finite resource)

We take again the model and explanations from [204] and [171] (see also the references therein) where empirical data specifying the parameters are given. A physiologically structured population equation is used again to describe algae as food (wit density $F(t)$ below) for small aquatic insects *Daphnia*, with more evolved biological phenomenology than in Section 1.2.2. We now denote by $n(t,x)$ the density of *Daphnia* with size x (here we mean a length),

$$
\begin{cases}
\frac{\partial}{\partial t}n(t,x) + \frac{\partial}{\partial x}\left[g(F,x)n(t,x)\right] + d(F,x)n(t,x) = 0, & x_b \le x < \infty, \\[2mm]
g(F,x_b)n(t,x_b) = \int_{x_b}^{x_m} b(F,x)n(t,x)dx, \\[2mm]
\frac{dF}{dt} = \Psi(F) - \int_{x_b}^{x_m} I(F,x)n(t,x)dx.
\end{cases}
\tag{4.54}
$$

For example, with some constants γ, $x_m, \dots$,

$$
g(F,x) = \gamma\left(x_m \frac{F}{F_h + F} - x\right)_+, \quad \text{growth rate,}
$$

$$
b(F,x) = \mathbf{1}_{\{x > x_j\}} r_m x^2 \frac{F}{F_h + F}, \quad \text{reproduction rate,}
$$

$$
d(F,x) = \mu, \quad \text{death rate,}
$$

$$
I(F,x) = \nu x^2 \frac{F}{F_h + F}, \quad \text{feeding rate,}
$$

$$
\Psi(F) = \alpha F\left(1 - \frac{F}{K}\right), \quad \text{autonomous algal dynamics,}
$$

or

$$
\Psi(F) = \alpha\left(1 - \frac{F}{K}\right).
$$

The first choice corresponds to a logistic growth for algae themselves developing with constant nutrients; the second choice corresponds to a constant inflow of fresh non-reproducing food with constant deterioration.

Several remarks explain the shape of these functions:

(i) The positive part in $g(F,x)$ is important and makes the analysis simpler. It is not always true based on biological evidence; sea-anemones and flatworms do shrink during food scarcity but highly organized animals like Daphnia generally do not, according to [171].

(ii) Consequently, there is a maximal possible size $x_M(F) = x_m \frac{F}{F_h+F} \le x_m$ and growth will stop after it is reached. This justifies considering that $g(F,x_m) = 0$ and $n(t,x) = 0$ for $x \ge x_m$.

(iii) Setting $d = b = 0$ and integrating by parts, we deduce that the mass evolves according to

$$\frac{d}{dt} \int_{x_b}^{\infty} x^3 n(t,x) dx = 3 \int_{x_b}^{\infty} x^2 g(F,x) n(t,x) dx,$$

$$3\gamma \int_{x_b}^{x_m F/(F_h+F)} [x_m x^2 \frac{F}{F_h+F} - x^3] n(t,x) dx.$$

The loose term $3\gamma x^3$ means that maintenance is proportional to the mass of the animal while the gain term $3\gamma x_m x^2 \frac{F}{F_h+F}$ means that food is absorbed proportionaly to the surface of the animal, a term that is balanced by $I(F,x)$.

We refer to [46] for an analysis of such nonlinear models. It is much more complicated than the linear renewal equation. Typically, steady states exist but are not always global attractors and periodic solutions may exist (see [177]).

4.3.2 Balance law for cell division with finite resources

The size structured model (4.16) can be extended to include an extra-cellular environment $S(t)$. An example is, cf. [125],

$$\begin{cases} \frac{\partial}{\partial t} n(t,x) + \frac{\partial}{\partial x} [r(x,S) n(t,x)] + [d(x) + B(x,S)] n(t,x) = \int b(y,x;S) n(t,y) dy, \\ \\ n(t,x=0) = 0, \\ \\ \frac{\partial}{\partial t} S(t,x) + S(t) = S_0 - \int r(x,S(t)) n(t,x) dx. \end{cases}$$

$$(4.55)$$

Here $r(x,S)$ denotes the growth rate, $B(x,S)$ the division rate, $b(x,y;S)$ the partition function normalized with

$$\int b(x,y,S) dy = 2B(x,S), \qquad \int y b(x,y,S) dy = x B(x,S).$$

As usual $n(t,x)$ represents the number density of cells with size x and $S(t)$ a nutrient shared by the total population (substrate).

Again this model expresses a 'conservation law' for the chemical $S(t)$ since we have

$$\frac{d}{dt} \left[\int x n(t,x) dx + S(t) \right] = S_0 - S(t) = \int d(x) n(t,x) dx.$$

Other nonlinearities may come in this type of equation. In [24] the author considers a cannibalistic model for fish farms. They derive a model where the size structure equation in (4.55) is coupled with the equation

$$r(x,S) = e^{ax} S(t,x), \qquad S(t,x) = \varphi(x) \star n(t,\cdot), \qquad x \in \mathbb{R},$$

(in fact the authors works on the variable x which is the log of the size of the fishes), $b = 0$ but death terms are also nonlinear with the nonlinearity $\psi \star n \; n$. The function φ, ψ are data.

4.3.3   A model for internet TCP connections

Transport equations very similar to those presented in 4.2 arise in many other applications. By curiosity, we present an example here. It is a model for internet TCP connections which we take from [13]. Here $n(t, w)$ is the density of windows of size w routed to a router, $q(t)$ length of the queue at the bottleneck buffer of the router.

$$
\begin{cases}
\frac{\partial}{\partial t} n(t, w) + (1 - k(t)) \frac{\partial}{\partial w} n(t, w) = k(t)[4w\, n(t, 2w) - w\, n(t, w)], \\[2ex]
n(t, w = 0) = 0, \\[2ex]
\frac{dq}{dt}(t) = \max\left(0, (1 - k(t)) \int_{w>0} w\, n(t, w) dw - L\right), \\[2ex]
0 \leq k(t) = F(q(t)) \leq 1.
\end{cases}
\tag{4.56}
$$

Typically, the given function $F : \mathbb{R}^+ \to [0, 1]$ satisfies $F(q) = 0$ for $q < \underline{Q} < 1$ (some given threshold), F increasing (and for instance $F(q) = 1$ for $q \geq \overline{Q}$). Depending on the queue length, it is decided either to increase continuously the size of the windows (q small, this is the transport term) or to divide it by a factor two (q large, this is the fragmentation term). On the other hand, the buffer treats the messages with a constant rate $L > 0$ and they arrive with the rate $(1 - k(t)) \int_{w>0} w\, n(t, w) dw$.

Not only the first equation is remarkably analogous to the equal mitosis equation (4.1), but the nonlinear term shares several features with those of Sections 4.3.1 and 4.3.2 (they arise through a quantity independent of x or w, saturation effects are present through the 'max').

4.3.4   Huxley's model and actin-myosin interaction

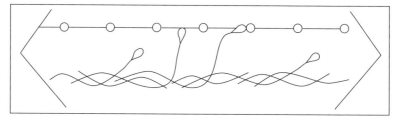

Figure 4.3: Principle of actin-myosin interaction.

A large class of muscles (called contractile or striated muscles, see [150]) work on the basis of sarcomeres (a kind of biomotor unit a few micrometers long), which all together lead to the biomechanical properties of muscles. At the individual level, sarcomeres are able to create a translation motion (contraction, relaxation of muscles) by electro-chemical power of myosin-actin bonds (see Figure 4.3). The (many) myosin heads are linked to a central myosin filament (backbone) but can move freely until they attach to the surrounding actin soft 'crystal type structure' at a distance ξ from a potential minimum of this actin structure. Then the myosin head is attracted towards the potential minimum by a force deriving from this potential where they have a tendency to detach by another chemical effect: ATP has filled the actin head and changed its electrical properties. A disymetry of the myosin potential (itself controlled by calcium ions Ca_{++}) creates an average motion in a given direction with a given velocity that is usually denoted $\dot{\varepsilon}$, the sarcomere velocity.

Huxley's model [141] aims to describe the density $\varrho(t,\xi)$ of myosin heads (also called cross-bridges) attached to the actin structure at a distance ξ from the potential center and creating a bulk motion of velocity $\dot{\varepsilon}$ (ε=strain) of the sarcomere. Huxley's model is given by

$$
\begin{cases}
\frac{\partial}{\partial t}\varrho(t,\xi) + \dot{\varepsilon}\frac{\partial}{\partial\xi}\varrho(t,\xi) = f(\xi)[1 - \varrho(t,\xi)] - g(\xi)\varrho(t,\xi), \\[2mm]
\varrho(t=0,\xi) = \varrho^0(\xi), \quad 0 \le \varrho^0(\xi) \le 1.
\end{cases}
\tag{4.57}
$$

The rate f describes how many free heads bind to the myosin structure (i.e. $1 - \varrho(t,\xi)$ after a normalization to unity of the density of free + attached heads). And g describes the unbinding rate of attached myosin heads becoming free after ATP hydrolysis. A recent review of the biophysical interpretations of such a model can be found in [193] as well as magnitude orders for these biomotors with respect to their size, number and forces created. Control problems are treated in [57].

One of the interesting points of such a model is the possibility to derive a macroscopic elastic model for the myofibre obtained by the collection of sarcomeres. This was achieved in [28] and uses the averaged quantities

$$
k_c(t) = \int_{\mathbb{R}} \varrho(t,\xi)d\xi, \quad (crossbridge \ density),
$$

$$
\sigma_c(t) = \int_{\mathbb{R}} \xi\varrho(t,\xi)d\xi, \quad (stress \ or \ tension).
$$

For the special case $f(\xi) + g(\xi) = Cst$ and integrating equation (4.57) yields

$$
\begin{cases}
\frac{d}{dt}k_c(t) = F - Ck_c(t), \\[2mm]
\frac{d}{dt}\sigma_c(t) - \dot{\varepsilon}k_c = G - C\sigma_c(t).
\end{cases}
\tag{4.58}
$$

On the other hand it can also be derived from a kinetic equation at the microscopic level

$$\tfrac{\partial}{\partial t}\psi(t,v,\xi) + v\tfrac{\partial}{\partial \xi}\psi(t,\xi) = f(\xi)[M(v) - \psi(t,v,\xi)] - g(\xi)\psi(t,v,\xi)$$
$$+[F_{ext} - \tfrac{\partial}{\partial \xi}V(\xi)]\,\tfrac{\partial}{\partial v}\psi(t,v,\xi) - \tfrac{\partial}{\partial v}[\tfrac{v-\dot\varepsilon}{T}\psi(t,v,\xi)].$$

$$(4.59)$$

Here $\psi(t,v,\xi)$ denotes the density of myosin heads attached to the actin structure at a distance ξ from the potential center and moving with the velocity v, and $M(v)$ denotes the (probability) density that a free actin site is reached with velocity v. The force created by the sawtooth potential V results in the third term in the right-hand side and the fourth (and last) one expresses the nearly solid structure of the sarcomere.

The nonlinearity comes from the self-consistent relaxation speed

$$\dot\varepsilon = \frac{1}{n(t)}\int_{\mathbb{R}\times\mathbb{R}} v\,\psi(t,v,\xi)\,dv\,d\xi,$$

with

$$n(t) = \int_{\mathbb{R}}\varrho(t,\xi)d\xi, \qquad \varrho(t,\xi) = \int_{\mathbb{R}}\psi(t,v,\xi)\,dv.$$

In the limit $T \to 0$ we expect that

$$\psi(t,v,\xi) \to \delta(v - \dot\varepsilon)\varrho(t,\xi),$$

and after v integration we recover Huxley's model.

There are two theories for the way this operates:

(i) The potential V is symmetric (around its minimum $\xi = 0$) and the motion comes from asymmetry in the rates $f(\xi)$ which has a peak at $\xi = \xi_0 > 0$ (the equilibrium position of free myosin heads) while $g(\xi)$ is symmetric (for instance).

(ii) The potential V is asymmetric (sawtooth) and the rates f, g are symmetric. Motion comes from the fact that attachment gives rise in average to a force always in the same direction.

In both cases the theorem should be: if F_{ext} is small enough then there is an average velocity $\dot\varepsilon$ in a given direction but to formalize this mathematically is an open question.

4.3.5 Molecular motors

Besides muscles that were treated in Section 4.3.4, molecular biomotors are numerous and based on motor proteins[1], [140]. They are at the origin of all the

---

[1]Motor proteins are molecular machines that directly convert chemical energy to work.

motions within the cell, which can be seen as a huge trafficking system (see [219] for the terminology and in a neuronal context) where molecular diffusion competes with oriented transport along various one-dimensional structures (see Figure 4.4 for an example). As one can find for instance in the book [140], recent progress of experimental devices has made it possible to measure forces at the molecular size, to visualize the deformation of filaments as microtubules or actin filaments, to evaluate how an enzyme can find a target. The next step has been to validate PDE or stochastic models for these phenomena and there are now many interesting mathematical questions arising from this description.

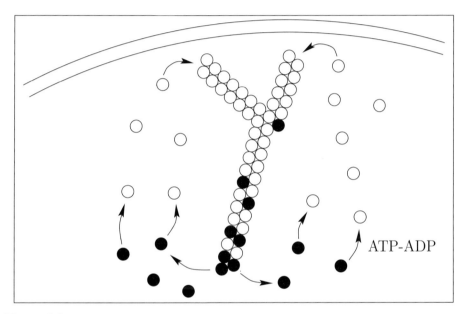

Figure 4.4: PRINCIPLE OF AN ACTIN FILAMENT PUSHING THE CELL MEMBRANE AND DRIVEN BY A MOLECULAR MOTOR.

4.3.6   Conventional kinesis

In several papers (see the review [146]), simple models for molecular biomotors have been derived where chemical energy is transformed to mechanical energy. The principle we consider here is that some molecule can reach two conformations (the density of each being denoted by n^1 and n^2 below). A bath of such molecules is moving in a filament and subject to two physical events. First, the filament induces a smooth, periodic and asymmetric potential seen differently by the two conformations (and denoted by $\psi^i(x)$, $i = 1$, 2 below). Second, fuel consumption triggers a conformational change between states 1 and 2 with rates denoted by $\nu^i > 0$ below. Being given that, at molecular scale, viscosity is important, this leads to the system of parabolic equations for the evolution of the densities $n^i(t, x)$:

$$\begin{cases} \frac{\partial}{\partial t} n^1 - \frac{\partial^2}{\partial x^2} n^1 - \frac{\partial}{\partial x}(\nabla\psi^1 \, n^1) + \nu^1 n^1 = \nu^2 n^2, & 0 \le x \le 1, \, t > 0, \\[2mm] \frac{\partial}{\partial t} n^2 - \frac{\partial^2}{\partial x^2} n^2 - \frac{\partial}{\partial x}(\nabla\psi^2 n^2) + \nu^2 n^2 = \nu^1 n^1, & \qquad\qquad (4.60) \\[2mm] \frac{\partial}{\partial x} n^i(x) - \nabla\psi^i \, n^i(x) = 0 \text{ at } x = 0, 1, & i = 1, 2. \end{cases}$$

Notice that the zero flux condition makes this system conservative,

$$\frac{d}{dt} \int_0^1 [n^1(t, x) + n^2(t, x)]dx = 0.$$

This can be interpreted in terms of our previous theory by noticing that in conservative cases the adjoint problem admits trivial solutions, here $\phi^1 = \phi^2 = Cst$, see (4.62) below.

This model, as well as several other biomotors, was analyzed in [61, 62, 152]. In particular, [61] proved that there is a positive steady state solution (N^1, N^2) that one can normalize by $\int_0^1 [N^1(x) + N^2(x)]dx = 1$.

$$\begin{cases} -\frac{\partial^2}{\partial x^2} N^1 - \frac{\partial}{\partial x}(\nabla\psi^1 \, N^1) + \nu^1 N^1 = \nu^2 N^2, & 0 \le x \le 1, \\[2mm] -\frac{\partial^2}{\partial x^2} N^2 - \frac{\partial}{\partial x}(\nabla\psi^2 N^2) + \nu^2 N^2 = \nu^1 N^1, & \qquad\qquad (4.61) \\[2mm] \frac{\partial}{\partial x} N^i(x) - \nabla\psi^i \, N^i(x) = 0 \text{ at } x = 0, 1, & i = 1, 2. \end{cases}$$

A simple way to see this goes again through the adjoint system, which is given by

$$\begin{cases} -\frac{\partial^2}{\partial x^2} \phi^1 + \nabla\psi^1 \, \frac{\partial}{\partial x}\phi^1 + \nu^1 \phi^1 = \nu^1 \phi^2, & 0 \le x \le 1, \\[2mm] -\frac{\partial^2}{\partial x^2} \phi^2 + \nabla\psi^2 \, \frac{\partial}{\partial x}\phi^2 + \nu^2 \phi^2 = \nu^2 \phi^1, & \qquad\qquad (4.62) \\[2mm] \frac{\partial}{\partial x} \phi^i(x) = 0 \text{ at } x = 0, 1, & i = 1, 2. \end{cases}$$

As already mentioned, it admits the trivial solution $\phi^1 = \phi^2 = Cst$, which proves that 0 is the first eigenvalue.

The very deep result in [61] is that the system exhibits a motor effect (the densities are higher near $x = 0$ than near $x = 1$ as depicted in Figure 4.5) under some precise asymmetry conditions on the potentials ψ^i and size conditions on the transition rates ν^i.

Our purpose here is less ambitious and is to give, without structure conditions on ψ^i and ν^i, an extension to this system of the Relative Entropy property (the structure behind is of course more general and relies on the coupling through

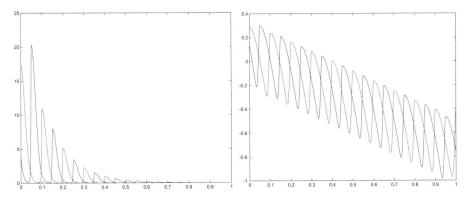

Figure 4.5: Motor effect for asymmetric potentials exhibited by the parabolic system (4.60). The figure represents the steady state given by (4.61) (left the densities themselves, right their logarithm). One can observe that the densities concentrate preferentially on the one hand as proved in [61].

zeroth order terms). It is not surprising that such systems also admit an entropy principle because they are positivity preserving. the method was previously used in [87] for another biomotor, the flashing rachet (the model includes a periodic potential and thus is related to Floquet Theory.

As a consequence we can study the solution of the parabolic system (4.60) and in particular prove the following properties of the solution.

Theorem 4.8. *Assume the potential ψ^i are smooth and $n^i(t = 0) \geq 0$ are integrable and bounded. Then,*

$$0 \leq n^i(t, x) \leq \max \left(\left\| \frac{n^1(t = 0, \cdot)}{N^1(\cdot)} \right\|_{L^\infty}, \left\| \frac{n^2(t = 0, \cdot)}{N^2(\cdot)} \right\|_{L^\infty} \right) N^i(x),$$

for all $x \in (0, 1)$, $i = 1, 2$. We define ρ by

$$\int_0^1 [n^1(t = 0, x) + n^2(t = 0, x)]dx = \rho \int_0^1 [N^1(x) + N^2(x)]dx,$$

then as $t \to \infty$, $\int_0^1 \left[|n^1(t, x) - \rho N^1(x)| + |n^2(t, x) - \rho N^2(x)| \right] dx$ decays to zero and

$$n^1(t, \cdot) \to \rho N^1(\cdot), \quad n^2(t, \cdot) \to \rho N^2(\cdot), \qquad in \ L^p(0, 1), \ \forall p \in [1, \infty[.$$

Again, the proof relies on a General Relative Entropy property of the parabolic equation (4.60) that can be expressed as

Lemma 4.6. *For all convex functions* $H : \mathbb{R} \to \mathbb{R}$, *and all solutions* (n^1, n^2) *to* *(4.60), we have the General Relative Entropy Inequality*

$$\frac{d}{dt}\int_0^1 [N^1(x)\, H\big(\tfrac{n^1(t,x)}{N^1(x)}\big) + N^2(x)\, H\big(\tfrac{n^2(t,x)}{N^2(x)}\big)]dx$$

$$= -\int \Big[N^1\, H''\big(\tfrac{n^1(t,x)}{N^1(x)}\big)[\tfrac{\partial}{\partial x}\big(\tfrac{n^1}{N^1}\big)]^2 + N^2\, H''\big(\tfrac{n^2(t,x)}{N^2(x)}\big)[\tfrac{\partial}{\partial x}\big(\tfrac{n^2}{N^2}\big)]^2 \Big]\,dx$$

$$- \int \nu^2 N^2 \Big[H'\big(\tfrac{n^1(t,x)}{N^1(x)}\big)[\tfrac{n^2(t,x)}{N^2(x)} - \tfrac{n^1(t,x)}{N^1(x)}] + H\big(\tfrac{n^1(t,x)}{N^1(x)}\big) - H\big(\tfrac{n^2(t,x)}{N^2(x)}\big) \Big]\,dx$$

$$- \int \nu^1 N^1 \Big[H'\big(\tfrac{n^2(t,x)}{N^2(x)}\big)[\tfrac{n^1(t,x)}{N^1(x)} - \tfrac{n^2(t,x)}{N^2(x)}] + H\big(\tfrac{n^2(t,x)}{N^2(x)}\big) - H\big(\tfrac{n^1(t,x)}{N^1(x)}\big) \Big]\,dx$$

$$\leq 0.$$

Proof. Since this computation follows exactly that of the similar principle for a parabolic equation (4.60), we just indicate again the main intermediary steps without details. We have

$$\frac{\partial}{\partial t}\big(\tfrac{n^1}{N^1}\big) - \frac{\partial^2}{\partial x^2}\big(\tfrac{n^1}{N^1}\big) + 2N^1 \frac{\partial}{\partial x}\big(\tfrac{n^1}{N^1}\big)\,\frac{\partial}{\partial x}\big(\tfrac{1}{N^1}\big) - \frac{\partial\psi^1}{\partial x}\frac{\partial}{\partial x}\big(\tfrac{n^1}{N^1}\big) = \nu^2 \tfrac{N^2}{N^1}[\tfrac{n^2}{N^2} - \tfrac{n^1}{N^1}].$$

Therefore, for any smooth function H, we arrive at

$$\frac{\partial}{\partial t}N^1 H\big(\tfrac{n^1}{N^1}\big) - \frac{\partial^2}{\partial x^2}N^1 H\big(\tfrac{n^1}{N^1}\big) + N^1 H''\big(\tfrac{n^1}{N^1}\big)\Big(\frac{\partial}{\partial x}\tfrac{n^1}{N^1}\Big)^2 - \frac{\partial}{\partial x}\big[\tfrac{\partial\psi^1}{\partial x}N^1 H\big(\tfrac{n^1}{N^1}\big)\big]$$

$$= \nu^2 N^2 H'\big(\tfrac{n^1}{N^1}\big)[\tfrac{n^2}{N^2} - \tfrac{n^1}{N^1}] + (\nu^2 N^2 - \nu^1 N^1)H\big(\tfrac{n^1}{N^1}\big).$$

After adding the similar result for the quantity $\frac{\partial}{\partial t}N^2 H\big(\tfrac{n^2}{N^2}\big)$ and integration in x, we arrive at the result. □

Proof of Theorem 4.8. Again these statements are direct consequences of the entropy inequality of Lemma 4.6, and we just indicate the choice of the entropy function $H(\cdot)$.

For the L^∞ bound we set (as in Section 6.3.1)

$$C = \max\left(\left\|\frac{n^1(t=0,\cdot)}{N^1(\cdot)}\right\|_{L^\infty}, \left\|\frac{n^2(t=0,\cdot)}{N^2(\cdot)}\right\|_{L^\infty}\right)$$

and the choice $H(u) = (u - C)_+$ concludes the argument. For the L^1 contraction principle it is enough to use $H(u) = |u|$. And the long time convergence again requires standard compactness arguments that have already been discussed in Chapter 3 and Section 4.2.4. □

Chapter 5

Cell motion and chemotaxis

5.1 How do cells move

Chemotaxis refers to several mechanisms through which cells can move in response to an external (usually chemical) signal. Several biological devices allow these small organisms to move. One of them is a family of "propellors" composed of flagella, very typical for instance of the prokaryotic [1] bacteria *Escherichia coli*, see Figure 5.1 left, which allow them to swim. Each flagella is activated by a 'biomotor' which responds to some chemical signal. This chemical signal can be directly emitted by the population of bacteria or created by a chemical reaction between the environment and the chemical emission of bacteria ([40]). Another example is *Proteus mirabilis* which can also swim, but in certain circumstances, can create multi-nucleoid hyperflagellated 'swarmers" (see [97]). More evolved living organisms like Amebae *Dictyostelium discoïdeum* crawl by sending forward an internal "arm" (an actin polymer as in Section 4.3.5), see Figure 5.3. These eukaryotes are also famous because they can aggregate under food restrictions. When the food resources are exhausted these amoebae emit the chemical signal, cyclic Adenosine Monophosphate (cAMP), which attracts the other amoebae and creates some kind of transition to a multicellular organism (fruiting body). Changes in the behavior of the entire body become determinant for survival of the population afterwards. Precise experiments on the way they react to chemoattractant can be found in Soll [212], with many questions on the possible interpretations of actual experiments which could have implications in terms of modeling. Finally let us mention *Myxobacteria* which have lived on soil surfaces for billion of years. They are equipped with two types of motors; they can send a long hair (pilus) in front of them, stick it into another bacterium, and retract it, thus pulling forward the

---

[1] In comparison to eukaryotes, in prokaryotes the chromatin (genetic material) is not enclosed inside the nucleus but is loosely arranged within the cytoplasm. Moreover there is no structure like a cytoskeleton to organize the displacement of the components within the cell.

bacterium; they can also "push" using similar devices ([147], Figure 5.1 right). This induces a movement which can also lead *Myxobacteria* to create fruiting bodies.

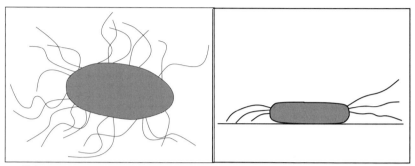

Figure 5.1: A REPRESENTATION OF BACTERIUM *Escherichia coli* AND ITS FLAGELLA.(LEFT) AND OF A *Myxobacterium* AND ITS PILUS (RIGHT).

Migration of cells occurs in another domain of biomedical science, namely cancer modeling. Development of cancers are very much related to the ability of cancerous cells to move, and thus spread and develop faster than healthy cells. Recently, authors have distinguished the so-called mesenchymal migration, in comparison to ameboid migration (see [111]), as represented in Figure 5.2. Mesenchymal migration is characterized by higher levels of adhesion on the extracellular matrix, of cell elongation, and by degradation of ECM. Another aspect of interaction between cell migration and cancer developments arises for solid tumors; they can secrete chemical signals (Vascular Endothelial Growth Factors) that produce the so-called angiogenesis effect. This refers to the formation of capillary network structure of blood vessels in the direction of the tumor; again the endothelial cells migrate towards the tumor by chemotactic motion.

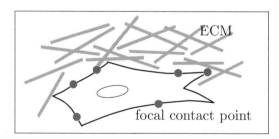

Figure 5.2: A REPRESENTATION OF A FIBROBLAST MOVING IN EXTRA-CELLULAR MATRIX ACCORDING TO [111].

Therefore we can understand two very extreme situations. Either the cells emit a (chemical) signal that attracts the other individuals (we refer to this as chemotaxis) or the external signal is consumed by the organism and we refer to this as angiogenesis model for the sake of simplicity. An intermediary modeling (in the sense that the chemoattractant is consumed but also produced indirectly) consists in the chemical synthesis of an exogeneous product and another substance exhausted by the bacteria ([58, 156, 166]).

These various biological issues and mathematical questions around cell motion can also be found in the Lecture Notes in Biomathematics edited by Alt and Hoffmannn [6], in the book of Murray (volume 2) [178] which also contains very complete information on this subject and in the survey papers [121] for colonies formation and [137] for mathematical aspects.

In this context, one of the places where mathematics plays a role is to understand how one can observe collective patterns from individual responses to a mean signal emitted by the cells themselves. These patterns are typically of two kinds:

(i) traveling waves are observed, for instance periodic swarm rings for *Proteus Mirabilis* ([97]) or band dynamics for *Escherichia coli* ([40] and the references therein),

(ii) aggregate formation in *E. coli* and *Dictyostelium*.

These questions have attracted very much attention among mathematicians and physicists. One reason is that an elementary modeling leads to a fascinating nonlinear model (Keller–Segel system) on which the mathematical literature is enormous. The survey by Horstmann [137, 138] gives the state of the art on these questions.

We present now these models which share the same mathematical structure. Namely the nonlinearity representing the attracting force arises as a first order transport term in a diffusion equation.

5.2　Chemotaxis: the Keller–Segel system $(d \geq 2)$

The simplest and most classical model used to describe the collective motion of cells has been developed by Patlak [194] and Keller–Segel [148, 149]. It consists in a system which describes the evolution of the density of bacteria $\varrho(t, x)$, $t \geq 0$, $x \in \mathbb{R}^d$ and the concentration $c(t, x)$ of the chemical attracting substance,

$$\begin{cases} \frac{\partial}{\partial t}\varrho - \Delta\varrho + \operatorname{div}(\varrho\chi\nabla c) = 0, \quad t \geq 0, \; x \in \mathbb{R}^d, \\[2mm] \alpha\frac{\partial}{\partial t}c - \Delta c + \tau\, c = \varrho, \\[2mm] \varrho(t = 0) = \varrho^0 \in L^\infty \cap L^1_+(\mathbb{R}^d). \end{cases} \tag{5.1}$$

The first equation just expresses the random (brownian) diffusion of the cells with a bias directed by the chemoattractant concentration with a sensitivity χ. The chemoattractant c is directly emitted by the cell, diffused on the substrate and $\tau^{-1/2}$ represents its activation length.

The notation L^1_+ means nonnegative integrable functions, and the parabolic equation on ϱ gives nonnegative solutions (as expected for the cell density)

$$\varrho(t, x) \geq 0, \qquad c(t, x) \geq 0. \tag{5.2}$$

Another property we will use is the conservation of the total number of cells

$$m^0 := \int_{\mathbb{R}^d} \varrho^0(x) \, dx = \int_{\mathbb{R}^d} \varrho(t, x) \, dx. \tag{5.3}$$

We have set the problem in a full space for the sake of simplicity (the situation is complicated enough without bounds). On bounded domains Ω, one uses the no-flux boundary conditions $\frac{\partial}{\partial \nu} \varrho = \frac{\partial}{\partial \nu} c = 0$ on $\partial \Omega$ denoting by ν the outward normal.

We will mostly consider the limit α, $\tau \to 0$ and fix the positive parameter $\chi > 0$ (for an optimal result on the parabolic problem see [66]). Then, the Laplace equation admits several solutions ([98]) and, of course, we consider the solution determined by the fundamental solution. We obtain ($\alpha = 0$, $\tau = 0$ in (5.1))

$$\begin{cases} \frac{\partial}{\partial t} \varrho - \Delta \varrho + \mathrm{div}(\varrho \chi \nabla c) = 0, \quad t \geq 0, \ x \in \mathbb{R}^d, \\[2mm] \nabla c = -\lambda_d \, \frac{x}{|x|^d} \star \varrho, \qquad \lambda_d = 1/(d|B_d|), \\[2mm] \varrho(t=0) = \varrho^0 \in L^\infty \cap L^1_+(\mathbb{R}^d). \end{cases} \tag{5.4}$$

In two dimensions the situation is somewhat misleading because the corresponding concentration $c = \lambda_2 \ln |x| \star n$ is not positive. This means that the situation has been oversimplified, which is obvious from the beginning in view of more realistic literature [164, 178].

This system has been widely studied since the 1980s ([59]) and a recent survey of the mathematical results is due to Horstmann, [137, 138]. To present roughly the reason why the mathematics of this system is so interesting, let us mention that in dimension 1 there are global smooth and unique solutions. In dimension 2, the Keller–Segel system is critical in L^1 (for the dimension d it is critical in $L^{d/2}$). This means that for small initial mass it has been proved (initially by [145]) that the system is well-posed globally in time (see also 5.3 below). But there is blow-up, i.e. the solution does not remain bounded, for large mass. Radially symmetric solutions, and the various types of blow-up, have been widely studied: the density ϱ concentrates as a Dirac mass [127, 225, 78, 39, 208]. Numerical evidence of the creation of Dirac masses in non-radial situations is shown in [168] as well as the evolution of Dirac concentrations for this numerical solution. Figure 5.4 presents a computation in a rectangle with zero flux boundary condition on ϱ and Dirichlet boundary condition $c = 0$.

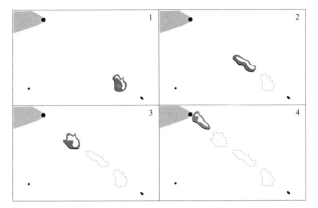

Figure 5.3: Motion of amoeba Dictyostelium discoïdeum in reaction to a chemoattractant emitted from the dark point at the upper left corner.

5.2.1 Free energy

The Keller–Segel system (5.1) admits several a priori estimates which reflects the basic modeling assumptions which have already been mentioned: the solution remains nonnegative and the total mass is conserved

$$\varrho(t, x) \geq 0, \qquad \int_{\mathbb{R}^d} \varrho(t, x) dx := m^0.$$

It also admits a dissipation principle for the free energy $\mathcal{E}_{cc}(t)$ which has been widely used ([35, 114, 181, 137, 138] …etc). In the case $\alpha = \tau = 0$, i.e., of equation (5.4) which is the only we treat here, it is defined as follows:

$$\begin{cases} \mathcal{E}_{cc}(t) = \int_{\mathbb{R}^d} [\varrho \ln(\varrho) - \frac{\chi}{2} \varrho c] \, dx, \\ \frac{d}{dt} \mathcal{E}_{cc}(t) = -\int_{\mathbb{R}^d} \varrho |\nabla \ln(\varrho) - \chi \nabla c|^2 \, dx \leq 0. \end{cases} \tag{5.5}$$

The main property of this energy is that it is composed of two terms; the entropy $\int_{\mathbb{R}^d} \varrho \ln(\varrho$ (which is in essence positive because small values of ϱ do not count in practice) and the (negative) potential energy $-\frac{\chi}{2} \int_{\mathbb{R}^d} \varrho c$. This competition on signs allows for a competition between the dissipative (diffusion) term and the drift.

To derive this free energy principle (5.5), we compute

$$\begin{aligned} \frac{d}{dt} \int_{\mathbb{R}^d} \varrho c \, dx &= \int_{\mathbb{R}^d} [c \frac{\partial}{\partial t} \varrho + \varrho \frac{\partial}{\partial t} c] \, dx \\ &= \int_{\mathbb{R}^d} [c \frac{\partial}{\partial t} \varrho - \Delta c \frac{\partial}{\partial t} c + \tau c \frac{\partial}{\partial t} c] \, dx \\ &= \int_{\mathbb{R}^d} [c \frac{\partial}{\partial t} \varrho - c \frac{\partial}{\partial t} \Delta c + \tau c \frac{\partial}{\partial t} c] dx \\ &= \int_{\mathbb{R}^d} [c \frac{\partial}{\partial t} \varrho + c \frac{\partial}{\partial t} \varrho] \, dx. \end{aligned}$$

Therefore

$$\frac{\chi}{2}\frac{d}{dt}\int_{\mathbb{R}^d}\varrho c\,dx = \chi\int_{\mathbb{R}^d}c\frac{\partial\varrho}{\partial t}\,dx$$
$$= \chi\int_{\mathbb{R}^d}c[\Delta\varrho - \operatorname{div}(\varrho\chi\nabla c)]\,dx$$
$$= \chi\int_{\mathbb{R}^d}\nabla c[-\nabla\varrho + \varrho\chi\nabla c].$$

But we also have,

$$\frac{d}{dt}\int_{\mathbb{R}^d}\varrho\ln(\varrho) = \int_{\mathbb{R}^d}(1+\ln(\varrho))[\Delta\varrho - \operatorname{div}(\varrho\chi\nabla c)]\,dx$$
$$= \int_{\mathbb{R}^d}\nabla\ln(\varrho)[-\nabla\varrho + \varrho\chi\nabla c].$$

Subtracting these two equalities, we recover (5.5).   □

The free-energy inequality is reminiscent of Poincaré and log-Sobolev inequalities (see Section 6.5, [12, 155, 228]) because it can be stated as

$$\frac{d}{dt}\int_{\mathbb{R}^d}e^{\chi c}u[\ln(u)+\frac{\chi}{2}c] = -4\int_{\mathbb{R}^d}e^c|\nabla\sqrt{u}|^2, \qquad u = \varrho e^{-\chi c}.$$

It is usual to look for an additional inequality for the dissipation rate in entropy inequalities. We let the reader check the following identity (for $\chi = 1$),

$$\frac{d}{dt}\int_{\mathbb{R}^d}e^c\left[|\nabla\sqrt{u}|^2 + \frac{\varrho^2}{2} - \frac{|\nabla c|^2}{2}\right] = -\sum_{i,j=1}^d\int_{\mathbb{R}^d}e^{\chi c}\left|\frac{\partial^2\sqrt{u}}{\partial x_i\partial x_j} - \frac{\partial\sqrt{u}}{\partial x_i}\frac{\partial\sqrt{u}}{\partial x_j}/\sqrt{u}\right|^2$$
$$+ \sum_{i,j=1}^d\int_{\mathbb{R}^d}e^c\frac{\partial^2 c}{\partial x_i\partial x_j}\frac{\partial\sqrt{u}}{\partial x_i}\frac{\partial\sqrt{u}}{\partial x_j}.$$

This free energy also appears in various other fields of application and has been studied first in the context of vortex systems in [44]. In particular, the 'critical mass' of chemotaxis appears in this paper as a 'critical temperature'.

Remark 5.1. The free energy for the full model (5.1) is given by

$$\begin{cases} \mathcal{E}_{cc}(t) = \int_{\mathbb{R}^d}\left[\varrho\ln(\varrho) - \chi\varrho c + \frac{\chi}{2}[|\nabla c|^2 + \tau c^2]\right]\,dx, \\ \frac{d}{dt}\mathcal{E}_{cc}(t) = -\int_{\mathbb{R}^d}\varrho|\nabla\ln(\varrho) - \chi\nabla c|^2\,dx - \frac{\chi\alpha}{\tau}\int_{\mathbb{R}^d}|\frac{\partial c}{\partial t}|^2dx \leq 0. \end{cases} \tag{5.6}$$

5.2.2   Existence for small $L^{d/2}$ norm ($d > 2$)

We begin with the general situation of the Keller–Segel system set in $\mathbb{R}^d$ for $d > 2$ and will come back to the dimension $d = 2$ later. We show the existence of solutions for small initial data. The natural scaling here leads us to consider the space $L^{d/2}$. For a theory in more elaborate spaces see [32].

Theorem 5.1. *Let $d > 2$. There is a universal constant $K(d)$ such that if*

$$\varrho^0 \in L^1(\mathbb{R}^d), \qquad \chi \, \|\varrho^0\|_{L^{d/2}(\mathbb{R}^d)} \leq K(d), \tag{5.7}$$

then the system (5.4) admits a global weak solution $\varrho \in L^\infty\big(\mathbb{R}^+; L^1 \cap L^{d/2}(\mathbb{R}^d)\big)$, and

$$
\begin{cases}
\|\varrho(t)\|_{L^{d/2}(\mathbb{R}^d)} \leq \|\varrho^0\|_{L^{d/2}(\mathbb{R}^d)}, & \varrho \in L^{1+d/2}(\mathbb{R}^+ \times \mathbb{R}^d), \\[2mm]
\nabla \varrho^{d/4} \in L^2(\mathbb{R}^+ \times \mathbb{R}^d), & \|\nabla c\|_{L^{d(d+2)/(d-2)}(\mathbb{R}^d)} \in L^{1+d/2}(\mathbb{R}^+),
\end{cases}
\tag{5.8}
$$

$$\nabla c(t) \in L^\infty(\mathbb{R}^+; L^q(\mathbb{R}^3)), \qquad \frac{d}{d-1} < q \leq d, \tag{5.9}$$

and,

$$
\begin{cases}
\|\varrho(t)\|_{L^{d/2}(\mathbb{R}^d)} \leq C(d)\, m^0\, t^{-\beta} & \text{for } t > 0,\ \beta = (d-2)/2, \\[2mm]
\|\varrho(t)\|_{L^p(\mathbb{R}^d)} \leq C(d)\, m^0\, t^{-\beta} & \text{for } t \geq T(d,p),\ \beta = \frac{d}{2}(1 - \frac{1}{p}),
\end{cases}
\tag{5.10}
$$

for $\varrho^0 \in L^p(\mathbb{R}^d)$, $p \in [1, \infty[$.

This theorem expresses that in the smallness regime, the quadratic term of the equation (namely $n\nabla c$) is so small that it does not have a qualitative effect. Indeed, the time decay stated here is exactly that of the heat equation for an initial data in $L^1(\mathbb{R}^d)$ because

$$\left\| n^0 \star \frac{1}{(2\pi t)^{d/2}} e^{-|x|^2/2t} \right\|_{L^p(\mathbb{R}^d)} \leq C \|n^0\|_{L^1(\mathbb{R}^d)}\, t^{-\beta}, \qquad \beta = \frac{d}{2}\Big(1 - \frac{1}{p}\Big).$$

In fact the last inequality of (5.10) can be improved to a regularizing effect like hypercontractivity, because one does not need the additional assumption $\varrho^0 \in L^p(\mathbb{R}^d)$, the mere $L^{d/2}$ norm is enough, see [66].

Many variants and other results follow from the proof below. It is possible to show that this weak solution has more regularity (in Sobolev spaces) when the initial data has. We do not prove it here and refer to [35] for instance.

Proof of Theorem 5.1. We begin with a formal a priori estimate, based on an argument due to [145], then we deduce the long time convergence to 0. Finally, we indicate how to make the argument rigorous.

(i) The theory is based on the following a priori estimate for solutions to (5.4). It is obtained after multiplying the equation by $p\,\varrho^{p-1}$.

$$
\begin{aligned}
\frac{d}{dt} \int_{\mathbb{R}^d} \varrho^p + 4\frac{p-1}{p} \int_{\mathbb{R}^d} |\nabla \varrho^{p/2}|^2 &= -(p-1)\chi \int_{\mathbb{R}^d} \nabla \varrho^p \cdot \nabla c \\
&= (p-1)\chi \int_{\mathbb{R}^d} \varrho^{p+1} \\
&\leq C(d,p)\chi \int_{\mathbb{R}^d} |\nabla \varrho^{p/2}|^2 \, \|\varrho\|_{L^{d/2}}
\end{aligned}
\tag{5.11}
$$

after using Gagliardo-Nirenberg-Sobolev inequality (6.34).

We now choose $p = d/2$, and this leads to the inequality (with $\bar{C}(d, \frac{d}{2}) = \frac{4p}{p-1}C(d, \frac{d}{2})$)

$$\frac{d}{dt} \int_{\mathbb{R}^d} \varrho^{d/2} \le 4\frac{d-2}{d}\left[\bar{C}(d, \tfrac{d}{2})\chi\|\varrho\|_{L^{d/2}} - 1\right] \int_{\mathbb{R}^d} |\nabla \varrho^{d/4}|^2.$$

So, whenever we have initially

$$\bar{C}(d, \tfrac{d}{2})\chi\|\varrho^0\|_{L^{d/2}} \le 1/2,$$

i.e. condition (5.7), then $\int_{\mathbb{R}^d} \varrho(t)^{d/2}$ decreases for all times $t \ge 0$ (because the above condition remains satisfied) and the a priori estimate (5.8) follows because after time integration between 0 and t we find

$$2\frac{d-2}{d} \int_0^t \int_{\mathbb{R}^d} |\nabla \varrho^{d/4}|^2 \le \int_{\mathbb{R}^d} (\varrho^0)^{d/2},$$

as $t \to \infty$ we obtain the L^2 estimate. The control of $\int_{\mathbb{R}^d} \varrho^{p+1}$ as in (5.11) gives the estimate on $\varrho^{1+d/2}$. Then, we apply the Young convolution inequality to the equation on ∇c in (5.4), (a) from the bound $\varrho \in L^{1+d/2}(\mathbb{R}^+ \times \mathbb{R}^d)$ it gives the last inequality in (5.8), (b) from the bound $\varrho \in L^\infty(\mathbb{R}^+; L^1 \cap L^{d/2}(\mathbb{R}^d))$ it gives directly (5.9).

(ii) To go further and prove that $\|\varrho\|_{L^{d/2}}$ vanishes, we use the Gagliardo–Nirenberg–Sobolev inequality (6.34) departing from the conclusion of step (i),

$$\begin{aligned}
\frac{d}{dt} \int_{\mathbb{R}^d} \varrho^{d/2} &\le -C \int_{\mathbb{R}^d} |\nabla \varrho^{d/4}|^2 \\
&\le -C \int \varrho^{1+d/2}/\|\varrho\|_{L^{d/2}} \\
&\le -C(m^0)^{(\theta-1)(1+d/2)/\theta}\|\varrho\|_{L^{d/2}}^{(1+d/2-\theta)/\theta}
\end{aligned}$$

after using the interpolation inequality $\|\varrho\|_{L^{d/2}} \le (m^0)^{1-\theta}\|\varrho\|_{L^{1+d/2}}^{\theta}$ with $\theta = 1 - \frac{4}{d^2}$.

We set $u(t) = \|\varrho(t)\|_{L^{d/2}}$, set $\alpha = (1+d/2)(1-\theta)/\theta > 0$ and thus arrive at

$$\frac{d}{2}u(t)^{d/2-1}\frac{d}{dt}u(t) \le -C(m^0)^{-\alpha}u(t)^{(1+d/2-\theta)/\theta},$$

and thus

$$u(t)^{(-1-d/2+d\theta/2)/\theta}\frac{d}{dt}u(t) \le -C(m^0)^{-\alpha},$$

which also leads to

$$\frac{1}{\alpha}\frac{d}{dt}u(t)^{-\alpha} \ge C(m^0)^{-\alpha}.$$

From the above differential inequality, we deduce that $\|\varrho(t)\|_{L^{d/2}} \le C\,m^0\,t^{1/\alpha}$ for large time. Therefore the time decay statement is proved with $\beta = 1/\alpha = (d-2)/2$.

(iii) We prove (5.10) coming back to (5.11). Since we have proved that $\|\varrho(t)\|_{L^{d/2}}$ decays to zero, for t larger than some $T(d,p)$ we have $\bar{C}(d,p)\|\varrho(t)\|_{L^{d/2}} \leq \frac{1}{2}$ and recover the same situation as in step (ii). We leave the corresponding calculation to the reader.

(iv) In order to prove the existence of a solution with the above properties, we consider an approximation of equation (5.4). We regularize the equation on c by a convolution

$$-\Delta c_\varepsilon = \varrho_\varepsilon \star \omega_\varepsilon, \qquad (5.12)$$

for some regularizing kernel $\omega_\varepsilon = \frac{1}{\varepsilon^d}\omega(\frac{x}{\varepsilon})$ with $\omega \in \mathcal{D}^+(\mathbb{R}^d)$, $\int_{\mathbb{R}^d} \omega = 1$. Then, following the standard parabolic theory, the system of the drift-diffusion equation on $(\varrho_\varepsilon, c_\varepsilon)$ admits a unique smooth solution, with fast decay in space, when the initial data is regularized and truncated. Then, we notice that the above a priori estimate still holds true since the only modification in the argument consists in the additional lines

$$
\begin{aligned}
\int_{\mathbb{R}^d} \nabla \varrho_\varepsilon^p \cdot \nabla c_\varepsilon &= -\int_{\mathbb{R}^d} \varrho_\varepsilon^p \Delta c_\varepsilon \\
&= \int_{\mathbb{R}^d} \varrho_\varepsilon^p \, \varrho_\varepsilon \star \omega_\varepsilon \\
&\leq \left(\int_{\mathbb{R}^d} \varrho_\varepsilon^{p+1}\right)^{p/(p+1)} \left(\int_{\mathbb{R}^d} (\varrho_\varepsilon \star \omega_\varepsilon)^{p+1}\right)^{1/(p+1)} \\
&\leq \int_{\mathbb{R}^d} \varrho_\varepsilon^{p+1}.
\end{aligned}
$$

As a consequence, the a priori bounds in the theorem hold true uniformly in ε and thus we can pass to the limit. This is easy because of the space gradient estimate that also follows from the a priori estimate, and because of the Lions–Aubin lemma which provides time compactness. In the limit, we obtain the results of Theorem 5.1. $\qquad\square$

5.2.3   Blow-up for large initial data ($d > 2$)

We continue with the dimension $d > 2$ and adapt an argument due to Nagai [179] to prove that blow-up occurs for large initial data. It was originally given in [31] and uses a smallness condition homogeneous to the $L^{d/2}$ norm although it is different (see the comment below).

Theorem 5.2. *For $d > 2$, we still set $m^0 := \int_{\mathbb{R}^d} \varrho^0(x)dx$. There is a constant C, small enough, such that when*

$$\chi \int_{\mathbb{R}^d} \frac{|x|^2}{2} \varrho^0(x)dx \leq C \, (\chi \, m^0)^{d/(d-2)},$$

then there is no global smooth solution, with enough decay in x at infinity, to the Keller–Segel system (5.4).

Also, notice that the assumption on the small $L^{d/2}(\mathbb{R}^d)$ norm for the existence Theorem 5.1 is in opposition with the smallness assumption in Theorem 5.2. To explain this we argue as follows. Let $R > 0$, then we have

$$
\begin{aligned}
m^0 &= \int_{\mathbb{R}^d} \varrho^0(x) dx \\
&\leq \int_{\{|x| \geq R\}} \varrho^0(x) dx + \int_{\{|x| \leq R\}} \varrho^0(x) dx \\
&\leq \int_{\{|x| \geq R\}} \frac{|x|^2}{R^2} \varrho^0(x) dx + C\, R^{d-2} \left[\int_{\{|x| \leq R\}} \varrho^0(x)^{d/2} dx \right]^{2/d} \\
&= \frac{2}{R^2} m_2^0 + C\, R^{d-2} \, \| \varrho^0 \|_{L^{d/2}(\mathbb{R}^d)} \\
&= C \| \varrho^0 \|_{L^{d/2}(\mathbb{R}^d)}^{2/d} \, (m_2^0)^{(d-2)/d},
\end{aligned}
$$

and the last line has been obtained with an appropriate choice of $R = m_2^0 / \| \varrho^0 \|_{L^{d/2}}$. Therefore the assumption that $(m_2^0)^{d-2}/(m^0)^d$ is small turns out to be incompatible with a small enough norm $\| \varrho^0 \|_{L^{d/2}(\mathbb{R}^d)}$.

Proof. We still denote $m_2(t) := \int_{\mathbb{R}^d} \frac{|x|^2}{2} \varrho(t,x) dx$ and use the formula

$$
\nabla c(t,x) = -\lambda_d \int_{\mathbb{R}^d} \frac{x-y}{|x-y|^d} \varrho(t,y) dy.
$$

Next, we compute using mass conservation,

$$
\begin{aligned}
\frac{d}{dt} m_2(t) &= d\, m^0 + \chi \int_{\mathbb{R}^d} \varrho(t,x)\, x \cdot \nabla c\, dx \\
&= d\, m^0 - \chi\, \lambda_d \int_{\mathbb{R}^d \times \mathbb{R}^d} \varrho(x) \varrho(y) \frac{x \cdot (x-y)}{|x-y|^d} \, dx\, dy \\
&= d\, m^0 - \frac{\chi\, \lambda_d}{2} \int_{\mathbb{R}^d \times \mathbb{R}^d} \varrho(x) \varrho(y) \frac{1}{|x-y|^{d-2}} \, dx\, dy
\end{aligned}
$$

after using the symmetry in the variables x and y in the last term. Then, we arrive at

$$
\begin{aligned}
\frac{d}{dt} m_2(t) &\leq d\, m^0 - \frac{\chi \lambda_d}{2R^{d-2}} \int_{|x-y| \leq R} \varrho(x) \varrho(y) \, dx\, dy \\
&\leq d\, m^0 - \frac{\chi \lambda_d}{2R^{d-2}} (m^0)^2 + \frac{\lambda_d}{2R^{d-2}} \int_{|x-y| \geq R} \varrho(x) \varrho(y) \, dx\, dy \\
&\leq d\, m^0 - \frac{\chi \lambda_d}{2R^{d-2}} (m^0)^2 + \frac{\chi \lambda_d}{2R^d} \int_{\mathbb{R}^d \times \mathbb{R}^d} \varrho(x) \varrho(y) |x-y|^2 \, dx\, dy \\
&\leq d\, m^0 - \frac{\chi \lambda_d}{2R^{d-2}} (m^0)^2 + \frac{2\, \chi\, \lambda_d}{R^d} m^0 \int_{\mathbb{R}^d} |x|^2 \varrho(x)\, dx \\
&\leq d\, m^0 - \frac{\chi \lambda_d}{2R^{d-2}} (m^0)^2 + \frac{4\, \chi\, \lambda_d}{R^d} m^0\, m_2(t).
\end{aligned}
$$

We choose $R = \mu\, (\chi\, m^0)^{1/(d-2)}$ with $\mu > 0$ small enough and we find

$$
\frac{d}{dt} m_2(t) \leq m^0 \left[-1 + C \frac{\chi\, m_2(t)}{(\chi\, m^0)^{d/(d-2)}} \right].
$$

When $\chi\, m_2(t=0)$ is small enough compared to $(\chi\, m^0)^d$, then $m_2(t)$ decreases for all times and thus we conclude

$$
\frac{d}{dt} m_2(t) \leq m^0 \left[-1 + C \frac{\chi\, m_2(0)}{(\chi\, m^0)^{d/(d-2)}} \right] < 0 \quad \text{(and is constant)}.
$$

This leads to a contradiction since $m_2(t)$ cannot vanish for smooth solutions. $\square$

5.3 Critical mass in dimension 2

In this section, we show that in dimension 2 the situation can be made more precise because there is an explicit critical mass, $m_{\text{crit}} = 8\pi/\chi$. We prove the global well-posedness in $\mathbb{R}^2$ for solutions with smaller mass and blow-up for larger mass (recall that mass is conserved according to (5.3)). The following result is proved in [88, 36] and similar thresholds are known in bounded domains ([114, 137, 138], see also [181] for the case of a parabolic equation on c).

The mathematical interest here is also to prove existence with an energy method rather than direct estimates based on Sobolev inequalities as in Section 5.2.2 and [145]. This strategy turns out to be much more accurate because the Gagliardo–Niremberg–Sobolev inequalities introduce thresholds on the critical mass that are not exact.

Theorem 5.3. *In* $\mathbb{R}^2$, *assume* $\int_{\mathbb{R}^2} |x|^2 \varrho^0(x) dx < \infty$.

(i) *(Blow-up) When the initial mass satisfies*

$$m^0 := \int_{\mathbb{R}^2} \varrho^0(x) dx > m_{\text{crit}} := 8\pi/\chi, \qquad (5.13)$$

then any solution to (5.4) becomes a singular measure in finite time.

(ii) *When the initial data satisfies* $\int_{\mathbb{R}^2} \varrho^0(x)|\log(\varrho^0(x))| dx < \infty$ *and*

$$m^0 := \int_{\mathbb{R}^2} \varrho^0(x) dx < m_{\text{crit}} := 8\pi/\chi, \qquad (5.14)$$

there are weak solutions to (5.4) satisfying the a priori estimates

$$\int_{\mathbb{R}^2} \varrho[|\ln(\varrho(t))| + |x|^2] \, dx \leq C(t),$$

$$\|\varrho(t)\|_{L^p(\mathbb{R}^2)} \leq C(p, t, \varrho^0) \qquad for \;\; \|\varrho^0\|_{L^p(\mathbb{R}^2)} < \infty, \quad 1 < p < \infty.$$

Remark 5.2. (i) A generic argument in [179] proves blow-up of solutions for $\Omega = B$, a ball under the conditions that the initial mass satisfies

$$\int_B \varrho^0(x) dx > 8\pi/\chi,$$

and that the second moment in x is also small enough. In more general bounded domains, still with no-flux boundary conditions, the critical mass is 4π because blow-up may occur on the boundary which intuitively acts as a reflection wall.

(ii) For radial solutions in $\mathbb{R}^2$, we will prove in Section 5.4 that one does not need a finite second moment for blow-up with a supercritical mass. A more subtle condition is however needed.

We also refer to Section 5.4 for more information on the blow-up (chemotactic collapse) in the case of radially symmetric solutions. For the case of more elaborate geometry our knowledge relies also on computations (see Figure 5.4 for instance).

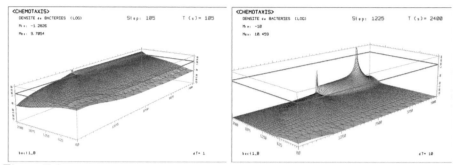

Figure 5.4: CHEMOTACTIC COLLAPSE FOR KELLER–SEGEL MODEL (5.4) SET IN A RECTANGLE. THIS IS A LOGARITHMIC SCALE AND FOR SYMMETRY REASONS, ONE FOURTH OF THE DOMAIN IS SHOWN. BLOW-UP IN FACT OCCURS IN THE LONG CENTRAL AXIS OF THE RECTANGLE. (COURTESY OF A. MARROCCO)

In the next subsections, we first give a proof of the blow-up statement (i) and then we consider existence (ii). More details and additional results (in particular on the long time behavior and intermediary asymptotic) can be found in [36]. Notice also that the above results extend to all dimensions for the log interaction kernel on c (see Section 5.5.4 and [48]).

5.3.1   Proof of blow-up, concept of weak solution

Formally the blow-up proof is very simple, and the difficulty here is to prove that the solution becomes a singular measure. We follow Nagai's argument, first assuming enough decay in x at infinity; afterwards we state a more precise result. It is based on the formula

$$\nabla c(t,x) = -\lambda_2 \int_{\mathbb{R}^2} \frac{x-y}{|x-y|^2} \varrho(t,y)dy, \qquad \lambda_2 = \frac{1}{2\pi}.$$

Then, we consider the second x moment

$$m_2(t) := \int_{\mathbb{R}^2} \frac{|x|^2}{2} \varrho(t,x)dx.$$

We have, from (5.4),

$$\begin{aligned}
\tfrac{d}{dt} m_2(t) &= \int_{\mathbb{R}^2} \tfrac{|x|^2}{2} [\Delta \varrho - \mathrm{div}(\varrho \chi \nabla c)]dx \\
&= \int_{\mathbb{R}^2} [2\varrho + \chi \varrho x \cdot \nabla c]dx \\
&= 2m^0 - \chi \lambda_2 \int_{\mathbb{R}^2 \times \mathbb{R}^2} \varrho(t,x)\varrho(t,y) \tfrac{x\cdot(x-y)}{|x-y|^2} \\
&= 2m^0 - \tfrac{\chi \lambda_2}{2} \int_{\mathbb{R}^2 \times \mathbb{R}^2} \varrho(t,x)\varrho(t,y) \tfrac{(x-y)\cdot(x-y)}{|x-y|^2}
\end{aligned}$$

(this last equality just follows by a symmetry argument, interchanging x and y in the integral). This yields finally,

$$\frac{d}{dt}m_2(t) = 2m^0(1 - \frac{\chi}{8\pi}m^0).$$

Therefore if we have $m^0 > 8\pi/\chi$, we arrive at the conclusion that $m_2(t)$ should become negative in finite time which is impossible since ϱ is nonnegative. Therefore the solution cannot be smooth until that time.

In order to go further, analyze more precisely this proof and prove the statement (i) of Theorem 5.3, we need a concept of *weak solution to the Keller–Segel system* that can handle L^1 solutions and that was used in [208]. To do that, we use the usual definition of solutions in distribution sense but take advantage of the symmetry in the drift term. Let $\psi \in \mathcal{D}(\mathbb{R}^2)$ be a test function, and test it in (5.4); we arrive at

$$\frac{d}{dt} \int_{\mathbb{R}^2} \psi(x)\varrho(t,x)$$
$$= \int_{\mathbb{R}^2} \Delta\psi(x)\varrho(t,x)dx - \frac{\chi}{2\pi} \int_{\mathbb{R}^2 \times \mathbb{R}^2} \nabla\psi(x).\frac{x-y}{|x-y|^2}\varrho(t,x)\varrho(t,y)dx\,dy.$$

In this equation we still need to make sense of the singularity of order $1/|x-y|$. This can be avoided in defining solutions as uniformly bounded measures in x, and weakly continuous in time, such that

$$\frac{d}{dt} \int_{\mathbb{R}^2} \psi(x)\varrho(t,x)$$
$$= \int_{\mathbb{R}^2} \Delta\psi(x)\varrho(t,x)dx - \frac{\chi}{4\pi} \int_{\mathbb{R}^2 \times \mathbb{R}^2} [\nabla\psi(x) - \nabla\psi(y)].\frac{x-y}{|x-y|^2}\varrho(t,x)\varrho(t,y)dx\,dy.$$
$$\tag{5.15}$$

Because $[\nabla\psi(x) - \nabla\psi(y)].\frac{x-y}{|x-y|^2}$ is bounded by 1, this definition of weak solutions makes perfect sense for $\varrho \in L^\infty(\mathbb{R}^+; L^1(\mathbb{R}^d))$.

Notice for instance that weak solutions are mass conservative. Indeed, we can choose a test function $\psi_R(x) = \psi(|x|/R)$ with ψ a smooth function such that $\psi(r) = 1$ for $r \le 1/2$, $\psi(r) = 0$ for $r \ge 1$. Then

$$\left| \int_{\mathbb{R}^2} \Delta\psi_R(x)\varrho(t,x)dx \right| \le \frac{C}{R^2} \int_{\mathbb{R}^2} \varrho(t,x)dx \xrightarrow[R\to\infty]{} 0,$$

$$\left| \int_{\mathbb{R}^2 \times \mathbb{R}^2} [\nabla\psi_R(x) - \nabla\psi_R(y)].\frac{x-y}{|x-y|^2}\varrho(t,x)\varrho(t,y)dx\,dy \right|$$
$$\le \frac{C}{R^2} \int_{\mathbb{R}^2 \times \mathbb{R}^2} \varrho(t,y)\varrho(t,x)dydx\,dy \xrightarrow[R\to\infty]{} 0.$$

Therefore, passing to the limit $R \to \infty$, we arrive (say test against a test function in time) at

$$\frac{d}{dt} \int_{\mathbb{R}^2} \varrho(t,x)dx = 0.$$

With the help of this concept of weak solution, we can also prove

Lemma 5.1. *A weak solution to (5.4) in the sense of (5.15) that satisfies*

$$\int_{\mathbb{R}^2} (1 + |x|^2) \varrho^0(x) \, dx < \infty$$

also satisfies, as long as it is an $L^1(\mathbb{R}^2)$ function,

$$\frac{d}{dt} \int_{\mathbb{R}^2} |x|^2 \, \varrho(t, x) \, dx = 4m^0 \left(1 - \frac{m^0}{m_{\text{crit}}} \right). \tag{5.16}$$

Remark 5.3. This concept of weak solution has only interest before the concentration as measure. Indeed, as proved in [208], the measure $n(x) = M\delta(x)$ is a weak solution in the sense of (5.15) (once correctly defined for measures, which is not direct), only for $M = \frac{4\pi}{\chi}$. Indeed, for that measure, (5.15) reduces formally (or for radially symmetric approximations of that measure) to

$$0 = M\Delta\psi(0) - \frac{M^2\chi}{4\pi} \sum_{ij=1}^{2} D_{ij}^2 \psi(0)\langle \frac{y_i y_j}{|y|^2}\rangle = M\Delta\psi(0) - \frac{M^2\chi}{8\pi}\Delta\psi(0).$$

Proof. Consider a family of functions $\psi_R(|x|) \in \mathcal{D}(\mathbb{R}^2)$ that grows nicely to $|x|^2$ as $R \to \infty$ as in the above argument for the total mass. Then we compute, as before,

$$\frac{d}{dt} \int_{\mathbb{R}^2} \psi_R \varrho \, dx = \int_{\mathbb{R}^2} \Delta\psi_R \varrho \, dx$$
$$- \frac{\chi}{4\pi} \int_{\mathbb{R}^2} \frac{(\nabla\psi_R(x) - \nabla\psi_R(y))\cdot(x-y)}{|x-y|^2} \varrho(t, x) \varrho(t, y) \, dx \, dy.$$

As before, both terms in the right-hand side are bounded (because $\Delta\psi_R$ and $\frac{(\nabla\psi_R(x) - \nabla\psi_R(y))\cdot(x-y)}{|x-y|^2}$ are bounded). Therefore the quantity $\int_{\mathbb{R}^2} \psi_R \varrho \, dx$ remains uniformly bounded and thus $\int_{\mathbb{R}^2} \psi_R \varrho \, dx < \infty$. Finally, as $R \to \infty$, we may pass to the limit in each term using the Lebesgue monotone convergence theorem (as long as we can dominate the various terms by the L^1 function ϱ). In this circumstance, we can pass to the limit and obtain equality (5.16). $\qquad\square$

This lemma concludes that $\varrho(t, x)$ cannot remain an L^1 function after the first time t^* when

$$\int_{\mathbb{R}^2} |x|^2 \varrho^0(x) \, dx + 4m^0 \left(1 - \frac{m^0}{m_{\text{crit}}} \right) t^* = 0,$$

and the statement (i) of Theorem 5.3 is proved. $\qquad\square$

For more about profile of solutions around isolated blow-up points, we refer to [127, 225], for the Hausdorff dimension of blow-up points (in the hyperbolic case), we refer to [78] and for dynamics of Dirac concentration points after blow-up, to [226].

5.3.2 Existence proof

We follow again [36] and decompose it in two steps. First, we prove equi-integrability and then prove the propagation of L^p norms. As in the higher dimensional case, we just derive a priori bounds on the solution and we skip the regularization argument which was already given in Section 5.2.2.

First step. Equi-integrability. We use the energy (see Section 5.2.1)

$$\mathcal{E}(t) = \int_{\mathbb{R}^2} \varrho \log \varrho \, dx + \tfrac{\chi}{4\pi} \int \int_{\mathbb{R}^2 \times \mathbb{R}^2} \varrho(t,x) \, \varrho(t,y) \, \log |x-y| \, dx \, dy,$$

and recall that $\mathcal{E}(t) \leq \mathcal{E}^0$.

We also recall the logarithmic Hardy–Littlewood–Sobolev inequality (see Section 6.5), for any function $f \geq 0$ with $\int_{\mathbb{R}^2} f = m^0$ and $\int_{\mathbb{R}^2} f |\ln f| < \infty$, then

$$\tfrac{m^0}{2} \int_{\mathbb{R}^2} f \log f \, dx + \int \int_{\mathbb{R}^2 \times \mathbb{R}^2} f(x) f(y) \log |x-y| \, dx \, dy \geq C(m^0).$$

Combining these two inequalities we find

$$\int_{\mathbb{R}^2} \varrho(t,x) \log \varrho(t,x) \, dx \leq \mathcal{E}_0 - \tfrac{\chi}{4\pi} \int \int_{\mathbb{R}^2 \times \mathbb{R}^2} \varrho(t,x) \, \varrho(t,y) \, \log |x-y| \, dx \, dy$$
$$\leq \mathcal{E}_0 - \tfrac{1}{4\pi} C(m^0) + \tfrac{m^0}{m_{\mathrm{crit}}} \int_{\mathbb{R}^2} \varrho(t,x) \log \varrho(t,x) \, dx.$$

For $\frac{m^0}{m_{\mathrm{crit}}} < 1$ this provides an a priori bound

$$\int_{\mathbb{R}^2} \varrho(t,x) \log \varrho(t,x) \, dx \leq C_1 := \tfrac{1}{1 - \frac{m^0}{m_{\mathrm{crit}}}} \left(\mathcal{E}_0 - \tfrac{1}{4\pi} C(m^0) \right). \qquad (5.17)$$

On the other hand the inequality (5.16) also provides a bound

$$\int_{\mathbb{R}^2} |x|^2 \varrho(t,x) \, dx \leq C_2(T), \qquad \forall t \leq T.$$

Combined with (5.17) it gives our final estimate for this step,

$$\int_{\mathbb{R}^2} \varrho(t,x) \big| \log \varrho(t,x) \big| \, dx \leq C_3(T). \qquad (5.18)$$

Indeed, we write

$$\int_{\varrho \leq 1} \varrho(t) \big| \log \varrho(t) \big| = \int_{\varrho \leq e^{-|x|^2}} \varrho(t) \big| \log \varrho(t) \big| + \int_{e^{-|x|^2} \leq \varrho \leq 1} \varrho(t) \big| \log \varrho(t) \big|$$
$$\leq \int |x|^2 e^{-|x|^2} + \int |x|^2 \varrho(t,x) dx \leq C + C \, t.$$

From the inequality (5.18) it is possible to deduce the existence of weak solutions (see [88]). Here we continue and prove L^p bounds.

Second step. L^p estimates. We use now Gagliardo–Nirenberg–Sobolev inequality (the constant is not optimal here)

$$\int_{\mathbb{R}^2} |u|^{2\gamma} \, dx \leq \gamma^2 \int_{\mathbb{R}^2} |u|^{2(\gamma-1)} \, dx \int_{\mathbb{R}^2} |\nabla u|^2 \, dx. \qquad (5.19)$$

Then, (5.4) yields, for $k > 1$,

$$\frac{d}{dt}\int_{\mathbb{R}^2}\frac{(\varrho-k)_+^p}{p(p-1)} \leq -\frac{4}{p^2}\int_{\mathbb{R}^2}|\nabla(\varrho-k)_+^{p/2}|^2\,dx + \chi\int_{\mathbb{R}^2}[\frac{(\varrho-k)_+^{p+1}}{p} + k(2p-1)\frac{(\varrho-k)_+^p}{p(p-1)}$$
$$+k^2\frac{(\varrho-k)_+^{p-1}}{p-1}]dx$$
$$\leq -\frac{4}{p^2}\int_{\mathbb{R}^2}|\nabla u|^2\,dx + \frac{\chi}{p}\int_{\mathbb{R}^2}u^{2\gamma}dx + C(k,p)\chi[m^0 + \int_{\mathbb{R}^2}(\varrho-k)_+^p\,dx]$$

after using Young's inequality with $u = (\varrho-k)_+^{p/2}$ and $\gamma = \frac{p+1}{p}$. Next we use (5.19) and for $k > 1$,

$$|u|^{2(\gamma-1)} = (\varrho-k)_+ \leq \varrho\frac{\ln_+(\varrho)}{\ln(k)} \leq \frac{C(t)}{\ln(k)}$$

with the constant $C_3(t)$ deduced from step 1. We arrive at

$$\frac{d}{dt}\int_{\mathbb{R}^2}(\varrho-k)_+^p\,dx \leq C(k,p)\chi[m^0 + \int_{\mathbb{R}^2}(\varrho-k)_+^p\,dx]$$
$$+\frac{1}{p^3}[(p+1)^2\frac{C_3(t)}{\ln(k)} - 4p]\int_{\mathbb{R}^2}|\nabla u|^2.$$

We can now conclude the proof. Being given $p > 1$ and $t > 0$, we choose k large enough so that the last term is negative and we conclude the proof of Theorem 5.3 (i).   □

Exercise. Consider the more general problem, say with $a > 0$ and $b > 0$,

$$\begin{cases} \frac{\partial}{\partial t}\varrho - \Delta\varrho + \mathrm{div}(\varrho^a\nabla c) = 0, & x \in \mathbb{R}^2, \\[2mm] -\Delta c = \varrho^b, & x \in \mathbb{R}^2, \\[2mm] \varrho(t=0) = \varrho^0 \in L^\infty \cap L_+^1(\mathbb{R}^2). \end{cases} \qquad (5.20)$$

Use a variant of the proof to show that,

(i) there is a global control of L^p norms whenever $a + b < 2$;

(ii) (ii) when $a + b = 2$, a smallness condition like (5.14) is needed.

See also [139] for higher dimensions.

5.4   Keller–Segel system: radially symmetric solutions

In fact [126, 127, 180] have made precise the type of blow-up that can occur in system (5.4) and the *chemotactic collapse* has been described (formally) for radial solutions in two dimensions. This means that, at the blow-up time, ϱ exhibits partial concentration as a Dirac mass of weight $\frac{8\pi}{\chi}$ at the origin plus an L^1 remainder. Notice also that by the argument above, we know that a Dirac concentration (at least lacking a singular measure) appears at blow-up time but it could be supported by a one-dimensional subset rather than a single point (but such a behavior is not expected generically).

In this section we show why radially symmetric solutions are simpler: the equation we derive satisfies a comparison principle. This fact has led to early results on this particular case ([179] and the references therein). We can also reach the steady state easily for $d = 2$ and have a new intuition of why the critical mass m_{crit} arises besides the energy and logarithmic Hardy-Littlewood-Sobolev inequality. We also prove a refined blow-up result (with a weaker condition than the second x-moment condition!) which is close enough to exhibit the shape of the chemotactic collapse solution. For complements and more recent papers on radially symmetric solutions we refer to [145, 33, 34].

5.4.1 Reduced system with radial symmetry

Radially symmetric solutions in d dimensions of the system (5.4) on $\varrho(t,r)$, $c(t,r)$, $r = |x|$, are equivalent to,

$$
\begin{cases}
\frac{\partial}{\partial t}(r^{d-1}\varrho) - (r^{d-1}\varrho')' + \chi(r^{d-1}\varrho c')' = 0, & t \geq 0, \quad r \geq 0, \\[2mm]
-(r^{d-1}c')' = r^{d-1}\varrho, \\[2mm]
\varrho'(t, r = 0) = c(t, r = 0) = 0,
\end{cases}
\tag{5.21}
$$

where $'$ stands for $\frac{\partial}{\partial r}$ and initial data have to be specified. And this can be reduced to a single equation on the number of cells contained in the ball of radius r and centered at the origin

$$
M(t,r) = 2\pi \int_0^r \sigma^{d-1}\varrho(t,\sigma)d\sigma = -2\pi r^{d-1}c'(t,r).
$$

We can integrate (5.21) and arrive at

$$
\begin{cases}
M'(r) = 2\pi r^{d-1}\varrho(r), \\[2mm]
\frac{\partial}{\partial t}M(t,r) - r^{d-1}\left(\frac{M'}{r^{d-1}}\right)' - \frac{\chi}{2\pi r^{d-1}}M'\,M = 0.
\end{cases}
\tag{5.22}
$$

Notice that the total mass is related to $M(t,r)$ by

$$
m^0 = \int_{\mathbb{R}^2}\varrho(t,x)dx = M(t, r = \infty),
$$

and that $M(t,r)$ should be increasing in r.

As mentioned earlier, there is a great simplification arising here. Namely, the Keller–Segel system is reduced to a single equation, which differs from Burgers equation (see [71, 196] for instance) only by its variable coefficients. In particular, we now have the comparison principle at hand!

5.4.2 Stationary states

In two dimensions, steady state solutions can be easily derived from this system by a further reduction. We write for time independent solutions

$$0 = r^2 (\frac{M'}{r})' - \frac{\chi}{2\pi} M' \, M = rM'' - M' + \frac{\chi}{4\pi}(M^2)'.$$

And thus, one more simple algebraic manipulation leads to

$$r\bar{M}' - 2\bar{M} + \frac{\chi}{4\pi}\bar{M}^2 = 0, \; r \geq 0, \qquad \bar{M}(0) = 0, \quad \bar{M}' \geq 0. \qquad (5.23)$$

The condition $\bar{M}(0) = 0$ just follows from the definition of $M(t, r)$ and is valid as long as there is no Dirac mass concentration at $x = 0$. It also provides us with the integration constant needed for the right-hand side of (5.23).

One readily checks that there are nontrivial solutions only in the case of a special value of the total mass, namely we have

Lemma 5.2. *There are nontrivial radial steady states to* (5.4) *only for the critical mass* $\bar{M}(\infty) = \frac{8\pi}{\chi} = m_{\mathrm{crit}}$, *given by*

$$\bar{M}_\lambda(r) = \frac{m_{\mathrm{crit}}}{1 + \lambda r^{-2}}.$$

This is a one-parameter family with $\lambda > 0$ related to the scale invariance of the Keller–Segel model (5.4). Recall also that the above value for the mass is equivalent to the threshold for blow-up in (5.13).

Coming back to the cell density thanks to the definition of $M(t, r)$, we obtain

$$\bar{\varrho}_\lambda(r) = \frac{M'(r)}{2\pi r} = \frac{\lambda \, m_{\mathrm{crit}}}{\pi} \frac{1}{(r^2 + \lambda)^2}. \qquad (5.24)$$

This function satisfies

$$\int_{\mathbb{R}^2} |x|^2 \bar{\varrho}_\lambda(|x|) dx = +\infty.$$

It has been studied in the context of vortex systems by [44] and one can readily check that this family minimized the energy functional (see Section 5.2.1). In particular it gives a hint on the process of formation of a Dirac mass in the Keller–Segel system because

$$\bar{\varrho}_\lambda(|x|) \xrightarrow[\lambda \to 0]{} m_{\mathrm{crit}} \, \delta(x = 0).$$

Remark 5.4. In a two-dimensional ball $B(0, R)$ there is a spherically symmetric solution given by the same formula (5.24) but with $0 \leq r \leq R$. Then the parameter λ serves to fix the total mass $m^0 < m_{\mathrm{crit}}$. Notice that this solution corresponds to zero flux on ϱ, but to $c = 0$ on $\partial B(0, R)$. The case of Neumann condition on c leads to multiple solutions (see [137, 138, 114, 49]).

5.4.3 Refined existence and blow-up

This solution may serve to give a hint on the behavior of generic radially symmetric solutions and in particular to understand why a singularity can arise only at the origin.

Theorem 5.4 (Global solutions). *In dimension 2, assume that we have*

$$m^0 = M(\infty) < m_{\mathrm{crit}}, \qquad M(t = 0) \le \bar{M}_{\lambda^0},$$

for some $\lambda^0 > 0$. Then the solution to (5.22) satisfies

$$M(t, r) \to 0 \quad as \ t \to \infty \quad locally \ uniformly,$$

and thus $\varrho(t)$ in (5.21) vanishes in $L^1(\mathbb{R}^2)$ locally.

Theorem 5.5 (Blow-up). *In dimension 2, assume that*

$$m^0 = M(\infty) > m_{\mathrm{crit}}, \qquad M(t = 0) \ge \bar{M}_{\lambda_0},$$

for some $\lambda^0 > 0$. Then the solution to (5.21) blows up in finite time.

This theorem expresses that, in the blow-up and existence results proved in Section 5.3, the limitation on the second moment are in fact useless in the radially symmetric case. They can be relaxed by the comparison condition with $\bar{M}_{\lambda^0}$, which we recall, has an infinite second momentum.

The proofs are also very different and use a comparison argument with sub- or supersolutions in the spirit of [145]. A related method can be found in [201] where the authors go further and prove the concentration of a mass $\frac{8\pi}{\chi}$ at the blow-up time, and after an appropriate extension of the notion of solutions, all the mass will concentrate at the origin.

Proof of Theorem 5.4. Being given, $M(\infty) < m_{\mathrm{crit}}$, we may always choose a positive number $\mu < 1$ which satisfies $\mu m_{\mathrm{crit}} > M(\infty)$. Then we consider the function

$$\bar{N}(t, r) = \min\left(M(\infty), \mu \, \bar{M}_{\lambda(t)}(r)\right), \qquad \lambda(t) = \lambda^0 + \frac{\chi}{\pi}(\mu^{-1} - 1)\left(\mu m_{\mathrm{crit}} - M(\infty)\right)t,$$

and we claim it is a supersolution to (5.22) (with $d = 2$). Indeed, we have

$$\frac{\partial \bar{N}}{\partial t} = -\bar{N} \, \frac{r^{-2}}{1 + \lambda r^{-2}} \, \frac{d\lambda(t)}{dt} \qquad \text{or } 0,$$

$$\bar{N}' = 2\bar{N} \, \frac{\lambda r^{-3}}{1 + \lambda r^{-2}} \qquad \text{or } 0.$$

Next, we may compute the radius $R(t)$ such that for $r > R(t)$ the minimum is attained by $M(\infty)$. It is given by

$$\lambda(t)R(t)^{-2} = \frac{\mu m_{\mathrm{crit}} - M(\infty)}{M(\infty)},$$

and for $r \le R(t)$, we have

$$\frac{\partial}{\partial t}\bar{N} - r(\frac{\bar{N}'}{r})' - \frac{\chi}{2\pi r}\bar{N}'\,\bar{N} = \bar{N}\,\frac{r^{-2}}{1+\lambda r^{-2}}[-\frac{d\lambda(t)}{dt} + \frac{\chi m_{\mathrm{crit}}}{\pi}(1-\mu)\frac{\lambda(t)r^{-2}}{1+\lambda r^{-2}}]$$

$$\ge \bar{N}\,\frac{r^{-2}}{1+\lambda r^{-2}}[-\frac{d\lambda(t)}{dt} + \frac{\chi m_{\mathrm{crit}}}{\pi}(1-\mu)\frac{\lambda(t)R(t)^{-2}}{1+\lambda R(t)^{-2}}]$$

$$= \bar{N}\,\frac{r^{-2}}{1+\lambda r^{-2}}[-\frac{d\lambda(t)}{dt} + \frac{\chi}{\pi}(\mu^{-1}-1)(\mu m_{\mathrm{crit}} - M(\infty))]$$

$$= 0.$$

But the minimum of two supersolutions is also a supersolution and thus $\bar{N}$ is indeed a supersolution to (5.22).

By the comparison principle, we deduce that the solution (5.22) satisfies $M(t,r) \le \bar{N}(t,r)$ (because it is satisfied initially). We conclude the proof just noticing that $R(t) \to \infty$ and for a given interval $r \in (0,R)$ we therefore have for t large enough $M(t,r) \le m_{\mathrm{crit}}/(1 + \lambda(t)R^2) \to 0$ as $t \to \infty$. □

Proof of Theorem 5.5. We follow the same lines as before and first choose, since $m_{\mathrm{crit}} < M(\infty)$, a value $\mu^0 > 1$ which satisfies $\mu^0 m_{\mathrm{crit}} < M(\infty)$. We consider now the function

$$\bar{N}(t,r) = \max\left(\frac{M(\infty)}{1 + \lambda^0 r^{-2}}, \frac{\mu^0 m_{\mathrm{crit}}}{1 + \lambda(t)r^{-2}}\right),$$

and we argue in two steps; (i) $\lambda(t) = \lambda^0 e^{-\alpha t}$ for $t < t_1$ large enough, (ii) $\lambda(t) = \lambda(t_1) - t + t_1$ for $t_1 < t < t_2 = t_1 + \lambda(t_1)$ the first time $\lambda(t)$ vanishes, and also we decrease slightly the value of μ during this step.

We will prove that $\bar{N}$ is a subsolution as it is the maximum of two subsolutions and this concludes the proof because we deduce that $M(t) \ge \bar{N}(t)$ and thus $M(t_2,0) > 0$ which is impossible for smooth solutions.

To prove that $\bar{N}$ is a subsolution, we first notice that $\frac{M(\infty)}{1+\lambda^0 r^{-2}}$ is a subsolution to (5.22) because it is an increasing function and $\frac{m_{\mathrm{crit}}}{1+\lambda^0 r^{-2}}$ is a solution. Secondly, we consider $t < t_1$ and we have to prove that $\frac{\mu^0 m_{\mathrm{crit}}}{1+\lambda(t)r^{-2}}$ is a subsolution only in an interval $r \le R(t)$ where it achieves the max, i.e.,

$$\frac{M(\infty)}{1 + \lambda^0 R(t)^{-2}} = \frac{\mu^0 m_{\mathrm{crit}}}{1 + \lambda(t)R(t)^{-2}}.$$

Notice that $r \le R(t) \le R_0$ with $\frac{M(\infty)}{1+\lambda^0 R_0^{-2}} = \mu^0 m_{\mathrm{crit}}$. Then, we compute

$$\frac{\partial}{\partial t}\bar{N} - r(\frac{\bar{N}'}{r})' - \frac{\chi}{r}\bar{N}'\,\bar{N} = \bar{N}\,\frac{r^{-2}}{1+\lambda r^{-2}}[-\frac{d\lambda(t)}{dt} + \frac{\chi m_{\mathrm{crit}}}{\pi}(1-\mu^0)\frac{\lambda r^{-2}}{1+\lambda r^{-2}}]$$

$$\le \bar{N}\,\frac{r^{-2}}{1+\lambda r^{-2}}[-\frac{d\lambda(t)}{dt} + \frac{\chi m_{\mathrm{crit}}}{\pi}(1-\mu^0)\frac{\lambda R_0^{-2}}{1+\lambda R_0^{-2}}]$$

$$\le \bar{N}\,\frac{r^{-2}}{1+\lambda r^{-2}}[-\frac{d\lambda(t)}{dt} + \frac{\chi m_{\mathrm{crit}}}{\pi}(1-\mu^0)\lambda R_0^{-2}]$$

$$= 0,$$

choosing

$$\frac{d\lambda(t)}{dt} = -\frac{\chi m_{\mathrm{crit}}}{\pi}(\mu^0 - 1)\lambda(t)R_0^{-2}, \qquad \alpha = \frac{\chi m_{\mathrm{crit}}}{\pi}(\mu^0 - 1)R_0^{-2}.$$

In a third step, we choose t_1 large enough so that $\lambda(t_1)$ is as small as necessary and then consider,

$$\lambda(t) = \lambda(t_1) - t + t_1, \qquad \mu(t) = \mu^0(1 - 2(t - t_1)),$$

for $t_1 < t < t_2$ with t_2 the time where $\lambda(t_2)$ vanishes, which can be chosen as close as we wish to t_1 by choosing t_1 large. Then we have (see the above calculation for the negative sign of the r-derivatives)

$$\frac{\partial}{\partial t}\bar{N} - r(\frac{\bar{N}'}{r})' - \frac{\chi}{r}\bar{N}' \; \bar{N} \leq \bar{N}[-\frac{1}{r^2+\lambda}\frac{d\lambda(t)}{dt} + \frac{d\mu(t)}{dt}/\mu(t)]$$
$$< 0,$$

as long as $t < t_2$ is close enough to t_1 which is exactly what we want. We have obtained again a subsolution which concludes the proof of Theorem 5.5. $\qquad\square$

5.5 Related chemotactic and angiogenetic systems

In this section we review some related models that have been proposed or studied recently; a more extensive presentation of many questions is in [137, 138].

5.5.1 Quorum sensing, volume filling, signal limiting

The blow-up phenomena is sometimes considered as a weakness for the Keller–Segel system (5.1). Therefore modifications have been proposed to prevent overcrowding and thus the blow-up phenomena. Examples have been proposed in [139, 130, 132, 154] and enter the general class

$$\begin{cases} \frac{\partial}{\partial t}\varrho - \mathrm{div}\big(D(\varrho)\nabla\varrho\big) + \mathrm{div}(\varrho\chi(\varrho)\nabla c) = 0, \\[2mm] \tau c - \Delta c = \varrho, \end{cases} \qquad (5.25)$$

with $\tau > 0$ a positive parameter and $D(u) \to \infty$ as $u \to \infty$ or $\chi(u) \to 0$ as $u \to \infty$. In such models, one can prevent blow-up using either the nonlinear diffusion $D(u)$ (volume filling), see [154], or nonlinear sensitivity $\chi(u)$ (quorum sensing) [139]. A very complete study of this kind of system based on energy methods can be found in [47]; the parabolic case is treated in [215] and the references therein. For similar questions at the kinetic level, see [55].

A natural question occurs now. We parametrize $D_n(u)$ and $\chi_n(u)$ in such a way that $D_n(u) \to 1$, $\chi_n(u) \to 1$ as $n \to \infty$. Then, one is supposed to recover

(5.1) which solutions may blow-up in finite time, thus providing a natural continuation of singular solutions after blow-up. Is this continuation unique? It has been observed numerically in [168] that this continuation in fact depends upon the truncation; a finite element method (that is some kind of truncation) usually exhibits steady concentration points but one can obtain 'concentration points' that move for the model with signal limiting

$$
\begin{cases}
\frac{\partial}{\partial t}\varrho - \Delta\varrho + \mathrm{div}(\varrho\chi\nabla c) = 0, \\[2ex]
\tau c - \Delta c = \varrho\, e^{-\varrho/A},
\end{cases}
\tag{5.26}
$$

where A is a large constant, as proposed in [29]. This phenomena is compatible with the formula proposed in [226] for the motion of the concentration points which indeed depends upon the truncation.

A special case of system (5.25) is

$$
\begin{cases}
\frac{\partial}{\partial t}\varrho - \mathrm{div}\big(D(\varrho)\nabla\varrho\big) + \mathrm{div}(\varrho(1-\eta\varrho)\chi\nabla c) = 0, \\[2ex]
\tau c - \Delta c = \varrho.
\end{cases}
\tag{5.27}
$$

This system satisfies clearly the maximum principle and as soon as $0 \le \varrho^0 \le \eta$, we have $0 \le \varrho(t) \le \eta$. Therefore solutions are global in time. However very interesting questions arise around various related asymptotic problems ([42, 86]) either for small diffusions or for $\eta \to 0$ because various intermediate asymptotic regimes occur.

Finally an interesting question remains to know if there is blow-up with sensitivity functions as $\chi(c) = \frac{\alpha}{1+c}$ in (5.1). Then an additional difficulty is that there is no known free energy functional. See the account on this subject in [200, 137].

5.5.2   Reacting chemoattractant

The system (5.1), and most of the variants, are unable to describe swarm rings or complex patterns that experiments can exhibit. Thus other ingredients are needed to see these traveling waves, spiral waves initiated patterns, multiple aggregation points. It can be proved that additional nonlinearities (sensitivity coefficient $\chi = \ln(c)$ for instance, birth and saturation terms) can lead to traveling waves [207, 37]. Attracting and repulsing substances, [23], can cause different patterns.

In [40], one can also find another biological explanation. The attractant (aspartate, $c(t, x)$ with previous notation) is produced by cells themselves when consuming fumarate ($f(t, x)$ below). The proposed model is [2]

$$
\begin{cases}
\frac{\partial}{\partial t}\varrho - \Delta\varrho + \mathrm{div}(\varrho \nabla c) = 0, \\[2mm]
-\Delta c = \varrho\, f, \\[2mm]
\frac{\partial}{\partial t} f - \beta \Delta f = -\gamma\, f\, \varrho.
\end{cases} \tag{5.28}
$$

The existence of solutions under scale invariant smallness assumptions is proved in [48]. It is an open question to prove blow-up or existence for large data.

5.5.3 Extra-cellular chemoattractant and active receptors

Still more elaborated models have been proposed in order to render complex patterns before blow-up. They include extra-cellular chemoattractant (u) and active receptors of the cells (v) and give rise to the system

$$
\begin{cases}
\frac{\partial}{\partial t}\varrho - \Delta\varrho + \mathrm{div}(\varrho \chi(v) \nabla u) = 0, \\[2mm]
\frac{\partial}{\partial t} u - \Delta u = F(\varrho, u, v), \\[2mm]
\frac{\partial}{\partial t} v = G(u, v).
\end{cases} \tag{5.29}
$$

Numerical simulations exhibiting patterns closely related to experimental observations can be found, see [164] for instance.

A general account for the biophysical assumptions (type of cells, type of substrate) sustaining such models can also be found in [178] as well as many references and various numerical simulations.

Notice that more accurate models at the kinetic level have been proposed recently. They take into account various internal states of the cells and can serve to derive macroscopic systems as (5.29). See [96, 85, 131] and the references therein.

5.5.4 The log interaction kernel and Modified Keller–Segel

From a mathematical point of view, the theory is much better when the chemoattractant c is derived from a different interaction kernel with a fractional diffusion law, namely

$$
\begin{cases}
\dfrac{\partial n}{\partial t} = \Delta n - \chi \nabla \cdot (n \nabla c) & t > 0,\ x \in \mathbb{R}^d, \\[2mm]
c = K_d * n & t > 0,\ x \in \mathbb{R}^d, \\[2mm]
n(t = 0) = n^0 \geq 0
\end{cases} \tag{5.30}
$$

---

[2] Compared to the system in [40], we have added a term f in the third equation of system (5.28) in order to impose a priori the inequality $f \geq 0$.

where the critical kernel K_d, and mass M_{crit} are defined as

$$K_d(z) = -\frac{1}{d\pi} \log |z|, \qquad M_{\text{crit}} = 2d^2\pi/\chi. \tag{5.31}$$

For this model, the Modified Keller–Segel system, introduced and studied in [49], it is proved that, for any dimension d, a theory holds that is very similar to that of two-dimensional Keller–Segel.

- For $\int_{\mathbb{R}^d} n^0(x)dx := M < M_{\text{crit}}$ there are global weak solutions, with regularizing effects in L^∞ (hypercontractivity).

- For $M > M_{\text{crit}}$ there is blow-up in finite time.

Behind this particular scaling that defines c, there are two properties that do not hold so nicely for the Keller–Segel system but do hold for the Modified Keller–Segel system; the second moment satisfies a closed equation and the energy contains exactly the terms of the Logarithmic Hardy–Littlewood–Sobolev inequality (see Section 6.5). These are the two equalities

$$\begin{aligned}
\frac{d}{dt} \int_{\mathbb{R}^d} \frac{1}{2}|x|^2 n(t,x)\, dx &= \int_{\mathbb{R}^d} \frac{1}{2}|x|^2 \nabla \cdot (\nabla n - \chi n \nabla c)\, dx \\
&= -\int_{\mathbb{R}^d} x \cdot (\nabla n - \chi n \nabla c)\, dx \\
&= \int_{\mathbb{R}^d} (\nabla \cdot x) n\, dx + \frac{\chi}{d\pi} \int_{\mathbb{R}^d \times \mathbb{R}^d} n(x) \frac{x \cdot (x-y)}{|x-y|^2} n(y)\, dy\, dx \\
&= M\left(d - \frac{\chi}{2d\pi} M\right)
\end{aligned} \tag{5.32}$$

and the energy functional

$$\mathcal{F}(n) = \int_{\mathbb{R}^d} n \log n - \frac{\chi}{2} \int_{\mathbb{R}^d} nc = \int_{\mathbb{R}^d} n \log n - \frac{\chi}{2} \int_{\mathbb{R}^d} n\, K * n, \tag{5.33}$$

satisfies the property $t \mapsto \mathcal{F}(n(t))$ is decreasing, more precisely

$$\frac{d}{dt} \mathcal{F}(n(t)) = -\int_{\mathbb{R}^d} n|\nabla \log n - \chi \nabla c|^2 \leq 0. \tag{5.34}$$

We refer to [49] for a study of the MKS in the full space and in bounded domains (where again the theory is much better).

5.5.5   Some numerical results

Specific patterns arise from a Keller–Segel model with growth terms presented in [178] where the reader can find the parameters and specific interpretation

$$\begin{cases}
\frac{\partial}{\partial t}\varrho - D\Delta\varrho + \operatorname{div}\left(\frac{\varrho}{(1+c^2)} \nabla c\right) + \mu\varrho(\delta - \varrho) = 0, \\[2mm]
\frac{\partial}{\partial t}\varrho - \Delta c = \beta\varrho^2 - c\varrho.
\end{cases} \tag{5.35}$$

Figure 5.5 presents numerical results exhibiting interesting patterns: from left to right, three different snapshots after the initialization with a high concentration at the upper right corner. The right most picture is the steady state.

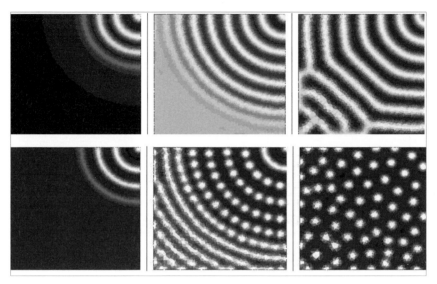

Figure 5.5: SOLUTIONS OF THE K.-S. SYSTEM WITH GROWTH (5.35). (UP) PARAMETERS HAVE BEEN CHOSEN IN THE TRANSITION RANGE BETWEEN DIFFUSION DOMINANT (NOT SHOWN) AND (DOWN) STABLE SPOTS LIKE STEADY STATES. (COURTESY OF A. MARROCCO)

5.5.6 A model from angiogenesis

Related models also appear in biomedical literature to describe the attraction of cells or blood vessels by an external signal (emitted by a tumor for instance). A process that is referred to as angiogenesis ([58, 107, 156, 166]). A simplified model, which shares some aspects of both previous models, is then

$$
\begin{cases}
\frac{\partial}{\partial t}\varrho - \Delta\varrho + \operatorname{div}(\varrho\nabla c) = 0, \\[2mm]
\frac{\partial}{\partial t}c = -\varrho\, c, \\[2mm]
\varrho^0 \in L^\infty \cap L^1_+(\mathbb{R}^d), \quad c^0 \in L^\infty_+(\mathbb{R}^d).
\end{cases}
\tag{5.36}
$$

This model admits an energy dissipation principle,

$$
\begin{cases}
\mathcal{E}_{ang}(t) = \int_{\mathbb{R}^d} [\varrho \ln(\varrho) + 2|\nabla\sqrt{c}|^2]\, dx, \\[2mm]
\frac{d}{dt}\mathcal{E}_{ang}(t) = -\int_{\mathbb{R}^d} \varrho\Big[|\nabla \ln(\varrho)|^2 + 6|\nabla\sqrt{c}|^2\Big] \le 0.
\end{cases}
\tag{5.37}
$$

Although apparently simpler than the Keller–Segel system, it presents also a deep mathematical interest. The first mathematical analysis goes back to [202] proving well posedness in one dimension and that, with some additional nonlinearities, the system can either blow-up or exhibit extinction, i.e. ϱ vanishes locally in finite time. As it is written here the model admits global weak solutions for initial data with finite energy. Strong solutions are obtained in ([67, 68]) but a smallness assumption is needed in dimensions higher than 3. It is an open question to prove L^p estimates for large initial data. We also refer to [107] for stability issues.

This angiogenesis model has an additional difficulty compared to the Keller–Segel system because there is no diffusion on c, which makes it of specific interest. On the other hand it describes repulsion rather than attraction and this simplifies some mathematical aspects, for instance both terms in the energy are positive.

5.6 Chemotaxis: kinetic equation

A kinetic model was proposed by [5, 189], based on detailed observations on the motion of *Escherichia coli*. As mentioned earlier (see Figure 5.1), bacterium *Escherichia Coli* is equipped with flagella. When rotated counterclockwise, the flagella act as a propellor resulting in a straight "run". When rotated clockwise the bundle of flagella separates, resulting in a "tumble" which reorients the cell but causes no significant change in location. *Bacillus Subtilis* (another bacterium) also moves by 'run and tumble' but the runs correspond to clockwise rotations and tumbles to counterclockwise rotation of the biomotor. During these runs, the bacterium can go as fast as $30\mu m/s$ (scaled to a car this would be close to $600km/h$!). As a consequence the bacterium moves along straight lines, suddenly stops to choose a new direction and then continues moving in this new direction. This phenomenon, called 'run and tumble', can be modeled by a stochastic process, the velocity-jump process [5, 189, 191, 214, 85, 192]. At the level of the population this is equivalent to writing a kinetic (or linear Boltzmann) type equation.

We first present the kinetic model in a first subsection, with the existence theory for the nonlinear model which was carried out in [54]. In second and third subsections, we explain the derivation of the 'macroscopic' Keller–Segel models. We refer to [119, 197, 227] for recent general presentations of the mathematical theories for kinetic equations. We also point out the recent kinetic model in [128] for mesenchymal motion i.e. cells moving in fiber networks (extra-cellular matrix) and degrading them.

5.6.1 Nonlinear transport-scattering equation

This 'run and tumble' process is very similar to that of *scattering* for neutrons that 'run' along straight lines until they encounter an atom and then are 'scattered' in a new direction ([72] Ch. 9, [197]). The governing equation is therefore reminiscent of the seminal Boltzmann equation and is posed for $t \geq 0$, $x \in \mathbb{R}^d$, $\xi \in V \subset \mathbb{R}^d$

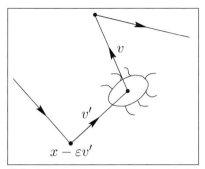

Figure 5.6: Run and tumble movement for E. Coli.

(it is natural to choose for instance V a ball of $\mathbb{R}^d$ centered at the origin, and we denote by $|V|$ the volume of V),

$$
\begin{cases}
\frac{\partial}{\partial t} f(t, x, \xi) + \xi \cdot \nabla_x f = \mathcal{K}(t, x, \xi), \\[2mm]
\mathcal{K}(t, x, \xi) = \int_V \left[K(c; \xi, \xi') f(t, x, \xi') - K(c; \xi', \xi) f(t, x, \xi) \right] d\xi', \qquad (5.38) \\[2mm]
f(t = 0, x, \xi) = f_0(x, \xi) \geq 0, \qquad f_0 \in L^1(\mathbb{R}^d \times V).
\end{cases}
$$

The gradient term expresses the transport of organisms with their own velocity ξ and the function $K(c; \xi, \xi') \geq 0$ is called the turning rate or scattering kernel and may also depend on (t, x) through a nonlocal dependency upon the chemoattractant concentration $c(t, x)$. It gives the rate $K(c; \xi, \xi')$ of organisms turning from velocity ξ' to ξ, and thus the rate $K(c; \xi', \xi)$ of organisms with velocity ξ that are subtracted from the balance on $f(t, x, \xi)$. Several possible forms for this kernel can be found in [129]; here, and the specific form of K is fundamental for the existence theory we develop below, we restrict ourselves to the case

$$
-\Delta c = n(t, x) := \int_V f(t, x, \xi) d\xi, \qquad (5.39)
$$

$$
K(c; \xi, \xi') = k_- \big(c(t, x - \varepsilon \xi') \big) + k_+ \big(c(t, x + \varepsilon \xi) \big). \qquad (5.40)
$$

The term k_- expresses roughly a delay $\varepsilon > 0$ in reaction time (a memory effect) which can be explained by the necessary saturation time of membrane receptors that control the rotation of the flagella. On the other hand, k_+ represents a possible knowledge of preferred directions (in the sense of higher chemoattractant concentration), see [5, 6, 129, 142] for further explanations. See [55] for a model preventing overcrowding.

Notice that these models do not suppose a comparison between two values of the concentration $c(t, x)$ at different locations, but only a biased turning rate according to the knowledge of a single value $c(t, x \pm \varepsilon \xi)$. We will assume that

$$
k_\pm \in C^1(\mathbb{R}^+; \mathbb{R}^+), \quad k_\pm(0) > 0, \quad 0 \leq k'_\pm \leq Q < \infty. \qquad (5.41)
$$

Also notice that, because of the diffusion equation on c, we arrive at a nonlinear *mean field equation* since the interaction is long range (see Subsection 5.8 for comments on this issue). This model shares many similarities with the gravitational Vlasov-Poisson equation (motion of self-attracting particles, [197]).

In terms of mathematical theory, this model is very interesting because of the lack of a priori estimates. The kernel being non-symmetric in ξ and ξ', we have at hand only the two properties

$$f(t,x,\xi) \geq 0 \quad \forall t \geq 0, \qquad \text{(minimum principle)}, \tag{5.42}$$

$$\int_{\mathbb{R}^d \times V} f(t,x,\xi) dx \, d\xi = \int_{\mathbb{R}^d \times V} f_0(x,\xi) dx \, d\xi, \qquad \text{(mass conservation)}. \tag{5.43}$$

This last property follows because we have, inverting the role of ξ and ξ' in the two integrals below (and using the Fubini theorem),

$$\int_V \mathcal{K}(t,x,\xi) d\xi = \int_V \int_V \left[K(c;\xi,\xi') f(t,x,\xi') - K(c;\xi',\xi) f(t,x,\xi) \right] d\xi' \, d\xi = 0.$$

Even in the linear case, to go further and derive L^p estimates on f for $p > 1$ is not so easy. The entropy structure behind such general models, which lacks a "detailed balance principle", also relies on the Generalized Relative Entropy method that we used for cell division models (see [176] and Section 6.4).

5.6.2   Existence theory

The existence theory for the nonlinear system (5.38), (5.40), (5.39) was settled in [54] (thus extending a result in [129] in the linear case) and yields

Theorem 5.6. *In dimension $d = 3$, assume that V is bounded, that (5.41) holds and that $f_0 \in L^\infty(\mathbb{R}^d \times V)$; then there is a unique solution to the system (5.38), (5.40), (5.39), $f \in C\big([0,\infty); L^1(\mathbb{R}^d \times V)\big)$, moreover we have the bounds*

$$0 \leq f(t,x,\xi) \leq C(t), \qquad 0 \leq n(t,x) \leq C(t) \, |V|,$$

$$\|c\|_{L^p(\mathbb{R}^d)} \leq C(t), \qquad d < p \leq \infty,$$

for some increasing constant $C(t)$.

This also holds in dimension $d = 2$ for the equation $-\Delta c + c = n$ for the chemoattractant.

This result provides global strong solutions and therefore shows a fundamental difference with the macroscopic model Patlak/Keller–Segel system since we have seen that the latter exhibits blow-up.

The method of proof of Theorem 5.6 has been extended in [142] to a remarkable result in dimension 3; namely we can use also, in place of the kernels in (5.40), kernels such as

$$K = k_-\big(\nabla c(t, x - \varepsilon \xi')\big), \quad \text{or} \quad K = k_+\big(\nabla c(t, x + \varepsilon \xi)\big),$$

but the sum is not allowed in this case when the dependency is on ∇c in place of c.

Theorem 5.7. *In any dimension $d > 2$, assume that V is bounded, that $f_0 \in L^\infty(\mathbb{R}^d \times V)$ and that the continuous kernel K satisfies either*

$$K(c; \xi, \xi') \leq K_0 + Q_+ c(x + \varepsilon \xi) + Q_+ |\nabla c(x + \varepsilon \xi)|,$$

or

$$K(c; \xi, \xi') \leq K_0 + Q_- c(x - \varepsilon \xi') + Q_- |\nabla c(x - \varepsilon \xi')|;$$

then there is a unique solution to the system (5.38), (5.39), $f \in C([0, \infty); L^1(\mathbb{R}^d \times V))$, moreover we have

$$0 \leq f(t, x, \xi) \leq C(t), \qquad 0 \leq n(t, x) \leq C(t) |V|,$$

$$\|c\|_{L^p(\mathbb{R}^d)} \leq C(t), \qquad \frac{d}{d-2} < p \leq \infty,$$

$$\|\nabla c\|_{L^p(\mathbb{R}^d)} \leq C(t), \qquad \frac{d}{d-1} < p \leq \infty$$

for some increasing constant $C(t)$.

This also holds in dimension $d = 2$ for the equation $-\Delta c + c = n$ for the chemoattractant and for the general dependency

$$K(c; \xi, \xi') \leq K_0 + Q_+ c(x + \varepsilon \xi) + Q_+ |\nabla c(x + \varepsilon \xi)| + Q_- c(x - \varepsilon \xi') + Q_- |\nabla c(x - \varepsilon \xi')|.$$

Extensions of these results have been obtained recently in [38].

Proof of Theorem 5.6. Again we only give the a priori estimates; they allow us to obtain the solutions after using a fixed point method. Also, for the proof we take $\varepsilon = 1$ in order to simplify notation.

We have,

$$\begin{cases} \frac{\partial}{\partial t} f(t, x, \xi) + \xi \cdot \nabla_x f & \leq \mathcal{K}_+(t, x, \xi), \\ \mathcal{K}_+(t, x, \xi) & = \int_V K(c; \xi, \xi') f(t, x, \xi') d\xi' \\ & = \int_V \left[k_- \big(c(t, x - \xi') \big) + k_+ \big(c(t, x + \xi) \big) \right] f(t, x, \xi') d\xi', \end{cases}$$

and thus, using Hölder inequality with $\frac{1}{p} + \frac{1}{p'} = 1$,

$$\frac{d}{dt} \int_{\mathbb{R}^d \times V} f^p(tx, \xi) dx \, d\xi = p \int_{\mathbb{R}^d \times V} f^{p-1} \mathcal{K}_+(t, x, \xi) dx \, d\xi$$

$$\leq p \left(\int_{\mathbb{R}^d \times V} f^{p'(p-1)} dx \, d\xi \right)^{1/p'} \left(\int_{\mathbb{R}^d \times V} \mathcal{K}_+(t, x, \xi)^p dx \, d\xi \right)^{1/p}.$$

Therefore, since $p'(p-1)=p$,

$$\tfrac{d}{dt}\int_{\mathbb{R}^d\times V} f^p(t,x,\xi)dx\,d\xi \le p\int_{\mathbb{R}^d\times V}\mathcal{K}_+(t,x,\xi)^p dx\,d\xi$$
$$\le C[\,\mathbf{I}+\mathbf{II}\,],$$

with

$$\mathbf{I} = \int_{\mathbb{R}^d\times V}\Big[\int_V k_-\big(c(t,x-\xi')\big)f(t,x,\xi')d\xi'\Big]^p dx\,d\xi$$
$$\le \int_{\mathbb{R}^d\times V}\Big[\int_V k_-\big(c(t,x-\xi')\big)^{p'}d\xi'\Big]^{p/p'}\int_V f^p(t,x,\xi')d\xi'\Big]dx\,d\xi$$
$$\le |V|\int_{\mathbb{R}^d\times V} f(t,x,\xi)^p dx\,d\xi\ \sup_{y\in\mathbb{R}^d}\Big[\int_V k_-\big(c(t,y+\eta)\big)^{p'}d\eta\Big]^{p/p'},$$

and

$$\mathbf{II} = \int_{\mathbb{R}^d\times V}\Big[\int_V k_+\big(c(t,x-\xi)\big)f(t,x,\xi')d\xi'\Big]^p dx\,d\xi$$
$$\le |V|^{p/p'}\int_{\mathbb{R}^d\times V} k_+\big(c(t,x-\xi)\big)^p\int_V f^p(t,x,\xi')d\xi'\,dx\,d\xi$$
$$\le |V|^{p/p'}\int_{\mathbb{R}^d\times V} f(t,x,\xi)^p dx\,d\xi\ \sup_{y\in\mathbb{R}^d}\Big[\int_V k_+\big(c(t,y+\eta)\big)^p d\eta\Big].$$

As a consequence, we arrive at the inequality

$$\tfrac{d}{dt}\int_{\mathbb{R}^d\times V} f^p(tx,\xi)dx\,d\xi \le C\int_{\mathbb{R}^d\times V} f(t,x,\xi)^p dx\,d\xi\, A[c(t,\cdot)], \qquad (5.44)$$

with, keeping in mind the assumption on $k_\pm$ given in (5.41),

$$A[c(t,\cdot)] = K_0 + Q\sup_{y\in\mathbb{R}^d}\big[\|c(t,y+\cdot)\|_{L^p(V)} + \|c(t,y+\cdot)\|_{L^{p'}(V)}\big]^p.$$

We can go one step further and write

$$c = c_L + c_S, \qquad c_S = \lambda_d\frac{\mathbf{I}_{\{|x|\le 1\}}}{|x|^{d-2}}\star n,$$

and the long range part interaction c_L is easy to handle since

$$0\le c_L(t,x) = \lambda_d\frac{\mathbf{I}_{\{|x|>1\}}}{|x|^{d-2}}\star n \le \lambda_d\|n\|_{L^1(\mathbb{R}^d)} \le \lambda_d\|f^0\|_{L^1(\mathbb{R}^d\times V)},$$

because of the mass conservation. Therefore the coefficient A arising in (5.44) is upper bounded by

$$A[c(t,\cdot)] \le C + Q\big[\|c_S(t,\cdot)\|_{L^p(\mathbb{R}^d)} + \|c_S(t,\cdot)\|_{L^{p'}(\mathbb{R}^d)}\big]^p.$$

The argument necessary to handle this short range part c_S is more elaborate because of the L^p and $L^{p'}$ norms.

To conclude we argue with a bootstrap argument taking into account that $\frac{\mathbf{I}_{\{|x|\le 1\}}}{|x|^{d-2}}\in L^q(\mathbb{R}^d)$ for all $1\le q<d/(d-2)$. We begin, as a first step, with $p_0=1$ and $n\in L^\infty\big(0,\infty);L^1(\mathbb{R}^d)\big)$ and thus, for all $q_0<d/(d-2)$,

$$\|c_S(t,\cdot)\|_{L^{q_0}(\mathbb{R}^d)} \le C\|n(t,\cdot)\|_{L^1(\mathbb{R}^d)} \le C\|f^0\|_{L^1(\mathbb{R}^d\times V)}.$$

With this range for the parameter q, we need to choose p in (5.44) so that $p \le q$ and $p' \le q$. This is possible with $p = q_0 > 2$ (close to 2 say) if $d/(d-2) > 2$, which holds for $d = 3$ (and for $d = 2$ if the equation on c is modified as in the statement so that the long range potential can be handled as before).

This means that the Gronwall inequality gives, for $p = p_1 = 2$,

$$\|f(t)\|_{L^p(\mathbb{R}^d \times V)} \le e^{Ct}\|f^0\|_{L^p(\mathbb{R}^d \times V)}, \qquad \|n(t)\|_{L^p(\mathbb{R}^d)} \le Ce^{Ct}\|f^0\|_{L^p(\mathbb{R}^d \times V)}.$$

Therefore, as a second step, using the Young inequality, $c_S \in L^\infty_{\text{loc}}(\mathbb{R}^+; L^{q_1}(\mathbb{R}^d))$ for all $q_1 \ge 1$ such that $1 + \frac{1}{q_1} > \frac{1}{2} + \frac{d-2}{d}$. Such q_1 are much larger than those q_0 of the first step. They allow us to choose much larger p_2 that satisfy both $p_2 \le q_1$ and $p_2' \le q_1$... and so on. The reader can easily convince himself that in three or four steps we arrive at $c \in L^\infty_{\text{loc}}(\mathbb{R}^+; L^\infty(\mathbb{R}^d))$, and the result on f is proved.

The result on c just follows by elliptic regularity. $\qquad \square$

Proof of Theorem 5.7. We now examine the case with dependency $k_\pm(\nabla c(x \pm \cdot))$ and skip the dependency on c, again for the sake of simplicity. The above computation still holds and leads to formula (5.44) with

$$A[c(t,\cdot)] = K_0 + \sup_{y \in \mathbb{R}^d} \left[Q_+\|\nabla c(t, y + \cdot)\|_{L^p(V)} + Q_-\|\nabla c(t, y + \cdot)\|_{L^{p'}(V)} \right]^p,$$

where we now distinguish $k_\pm$ with a Lipschitz constant $Q_\pm$.

In order to apply the method developed above, we now use a decomposition for the convolution kernel associated to ∇c in long and short ranges, namely,

$$\nabla c = \nabla c_L + \nabla c_S, \qquad \nabla c_S = \lambda_d \frac{x\, \mathbf{I}_{\{|x| \le 1\}}}{|x|^d} \star n,$$

and the long range interaction is still easy to handle

$$\nabla c_L = \lambda_d \frac{x\, \mathbf{I}_{\{|x| > 1\}}}{|x|^d} \star n, \qquad |\nabla c_L| \le C\|f^0\|_{L^1(\mathbb{R}^d \times V)}.$$

As before, we end up with the simpler estimate

$$A[c(t,\cdot)] = C + \left[Q_+\|\nabla c_S(t,\cdot)\|_{L^p(V)} + Q_-\|\nabla c_S(t,\cdot)\|_{L^{p'}(V)} \right]^p.$$

We explain now why only Q_+ or Q_- can be present in the turning kernel but not both. As before the idea is to bootstrap departing from $n \in L^\infty\big((0,\infty); L^1(\mathbb{R}^d)\big)$ and thus, for all $q_0 < d/(d-1)$,

$$\|\nabla c_S(t,\cdot)\|_{L^{q_0}(\mathbb{R}^d)} \le C\|n(t,\cdot)\|_{L^1(\mathbb{R}^d)} \le C\|f^0\|_{L^1(\mathbb{R}^d \times V)}.$$

But it turns out that for $d \ge 2$ we always have $d/(d-1) \le 2$. Therefore we cannot have both $p > q_0$ and $p' > q_0$. If only one term k_+ or k_- is present, say k_+ for instance, we can still continue. We use $p = q_0$, and

$$\|f(t)\|_{L^p(\mathbb{R}^d \times V)} \le e^{Ct}\|f^0\|_{L^p(\mathbb{R}^d \times V)}, \qquad \|n(t)\|_{L^p(\mathbb{R}^d)} \le Ce^{Ct}\|f^0\|_{L^p(\mathbb{R}^d \times V)}.$$

Therefore the Young inequality gives $n \in L^\infty_{loc}\big(\mathbb{R}^+; L^{q_1}(\mathbb{R}^d)\big)$ with $1+\frac{1}{q_1} > \frac{1}{q_0}+\frac{d-1}{d}$. Clearly one can choose any q_1 such that $\frac{1}{q_1} > 2\frac{d-1}{d} - 1 = \frac{d-2}{d}$ and thus we may choose again $p = q_1 > q_0$ and so on. Again we do not check the details of numerology leading to the L^∞ regularity.

If now k_- only is present, it is still simpler because we can directly choose $p = +\infty$, $p' = 1$ and we conclude. Indeed, $\frac{x\, 1_{\{|x|\le 1\}}}{|x|^d} \in L^1(\mathbb{R}^d)$ and thus

$$\|\nabla c_S(t,\cdot)\|_{L^1(\mathbb{R}^d)} \le C\|n(t,\cdot)\|_{L^1(\mathbb{R}^d)} \le C\|f^0\|_{L^1(\mathbb{R}^d\times V)}.$$

In dimension $d = 2$, with the equation $-\Delta c + c = n$, the method developed above still does not apply directly to the case

$$K(c;\xi,\xi') \le K_0 + Q_+ c(x+\varepsilon\xi) + Q_+|\nabla c(x+\varepsilon\xi)| + Q_- c(x-\varepsilon\xi') + Q_-|\nabla c(x-\varepsilon\xi')|.$$

An additional, and more refined, logarithmic loss in a limiting Young inequality is needed.

$$\|\nabla c\|_{L^2(\mathbb{R}^2)} \le C\|n(t,\cdot)\|_{L^1(\mathbb{R}^2)} \big(\log \|n(t,\cdot)\|_{L^2(\mathbb{R}^2)}\big)^{1/2}.$$

The proof is given in Lemma 6.9 in Appendix 6.5.

This inequality induces a logarithmic loss in the Gronwall argument. We do not carry out the details and refer to [142] for the end of the proof of Theorem 5.7 in the case $d = 2$. $\qquad\square$

5.7   Diffusion limit; back to the Keller–Segel equation

The classical field of application of kinetic equations consists in finding the so-called transport coefficients in macroscopic equations. The most remarkable example is to derive the pressure laws, viscosities and heat conductivities in the Navier–Stokes system from the Boltzmann equation derived in the early twentieth century by Chapman and Enskog (see [53] for instance). In the case at hand, this means that, departing from some knowledge, or some intuition, of the individual response of cells to the chemical stimulus described by the kernel $\mathcal{K}$, one can recover the coefficients of a macroscopic equation.

In order to achieve such a program, we need to find regimes where interactions reduce the kinetic equation in (t,x,ξ) to a macroscopic model with variables (t,x) only. In other words, we need to isolate small parameters which are related to the relaxation of the density $f(t,x,\xi)$ to specific dependency on ξ, called an equilibrium state. For the above mentioned reduction from Boltzmann equation to Navier–Stokes system, the Maxwellian distribution arises naturally

$$f(t,x,\xi) \approx \frac{\varrho}{(2\pi T)^{d/2}} e^{-|\xi-u|^2/(2T)}.$$

Other specific distributions arise in various fields of applications, such as plasma physics, with each time a direct physical interpretation, [75, 167, 196, 122] and Section 5.8.

There are several ways to perform this asymptotic analysis and to choose small parameters in kinetic equations. The most standard ones are the so-called hydrodynamic limits corresponding to the hyperbolic scale which we explain in the next section, and the diffusion limit corresponding to the parabolic scale which we perform now.

5.7.1 Rescaled kinetic chemotaxis model

The simplest example of a diffusion limit is to derive the heat equation from a scattering equation (see [15, 72, 192] and the references therein). The regime of interest is when the scattering operator dominates transport and this leads, in the average, to many 'tumbles', and thus to small macroscopic (averaged) velocities. Then a rescaling is introduced and here it is natural to use the small time scale ε arising in (5.40). The parabolic scale consists, after a change of time and space scales, in replacing the equation (5.38) by

$$
\begin{cases}
\frac{\partial}{\partial t} f_\varepsilon(t,x,\xi) + \frac{\xi}{\varepsilon} \cdot \nabla_x f_\varepsilon \\
\qquad + \frac{1}{\varepsilon^2} \int_V \big[K_\varepsilon(c_\varepsilon; \xi', \xi) f_\varepsilon(t,x,\xi) - K_\varepsilon(c_\varepsilon; \xi, \xi') f_\varepsilon(t,x,\xi') \big] d\xi' = 0, \\
f_\varepsilon(t=0,x,\xi) = f_0(x,\xi) \geq 0, \qquad f_0 \in L^1 \cap L^\infty(\mathbb{R}^d \times V).
\end{cases}
$$
(5.45)

The notation K_ε has only been used to put in evidence the dependency upon ε in the definition of K in (5.40) and we still consider that the chemoattractant concentration $c_\varepsilon(t,x)$ is related to f_ε through the Laplace equation (5.39), i.e.,

$$
-\Delta c_\varepsilon(x) = n_\varepsilon(x), \qquad x \in \mathbb{R}^d.
$$
(5.46)

The diffusion limit consists in studying the limit as ε vanishes. This nonlinear model (5.38), (5.40), (5.39), has been introduced and studied by several authors, [5, 189, 190, 191, 194, 129], and the linear case, when the chemoattractant concentration is a given smooth function, can be found in these references. The main originality, already present in the linear case, is to raise a drift term in the limit and not a mere diffusion as usual. A similar problem arises in other applications such as plasma physics, [75, 185].

Coming back to the question of transport coefficients, the outcome is to make the relation between two coefficients. At the microscopic scale we use the function k_+ for instance, and at the macroscopic scale we have the diffusion $D(c,n)$ and sensitivity $\chi(c,n)$ in the general Keller–Segel model

$$
\begin{cases}
\frac{\partial}{\partial t} n(t,x) - \mathrm{div}(D \cdot \nabla n) + \mathrm{div}(n\,\chi \cdot \nabla c) = 0, \quad t \geq 0, \ x \in \mathbb{R}^d, \\
\qquad\qquad\qquad\qquad -\Delta c = n.
\end{cases}
$$
(5.47)

A complete proof of convergence in the nonlinear case is given in [54]. Namely the follwing is proved.

Theorem 5.8. *Under the assumption* (5.40), $\int_V \xi \, d\xi = 0$, $|V| = 1$ *to simplify, there is a time* $T^* > 0$ *such that, as* $\varepsilon \to 0$,

$$f_\varepsilon(t, x, \xi) \rightharpoonup n(t, x) \, \mathbf{1}_{\{\xi \in V\}}, \quad \text{weakly in } L^2([0, T^*] \times \mathbb{R}^d \times V),$$

$$c_\varepsilon(t, x) \to c(t, x), \qquad \nabla c_\varepsilon(t, x) \to \nabla c(t, x),$$

and the general Keller–Segel system (5.47) *holds in the limit with initial data* $n^0(x) = \frac{1}{|V|} \int_V f_0(x, \xi) d\xi$ *and with the nonnegative symmetric diffusion matrix* D *and symmetric sensitivity matrix given by*

$$D(c) = \frac{1}{k_+(c) + k_-(c)} \int_V \xi \otimes \xi \, d\xi, \qquad \chi(c) = \frac{k'_+(c) + k'_-(c)}{k_+(c) + k_-(c)} \int_V \xi \otimes \xi \, d\xi.$$

This result expresses an interesting effect. The diffusion matrix $D(c)$ only arises from the symmetric part of the turning kernel K (at zeroth order), while the drift (and thus the sensitivity $\chi(c)$) arises from the antisymmetric part at first order in ε. In other words the memory effect is fundamental in order to obtain the observed collective movement of the cells leading to their aggregation.

We do not prove Theorem 5.8 but instead explain the importance of the symmetric and small antisymmetric parts. To do that we consider equation (5.45) with a turning kernel that satisfies

$$K_\varepsilon(c, ; \xi, \xi') = K_s(c; \xi, \xi') + \varepsilon K_{r,\varepsilon}(c; \xi, \xi'), \qquad K_s(c; \xi, \xi') = K_s(c; \xi', \xi) \geq k_m > 0,$$
$$\tag{5.48}$$

for some symmetric part K_s and some remainder (not necessarily antisymmetric at this stage) $K_{r,\varepsilon}$ which is however assumed to be bounded. For instance, consider the case where K_ε depends upon c as in (5.40), with a smooth function c, then we might choose

$$K_s = k_-(c) + k_+(c).$$

5.7.2   Uniform estimate

A first way to see the necessity of the decomposition (5.48), is the derivation of estimates on f_ε uniform in ε. This is expressed in

Proposition 5.1. *With the assumption* (5.48), *the solutions to* (5.45) *satisfy for all* $t \leq T$,

$$\|f_\varepsilon(t)\|_{L^2(\mathbb{R}^d \times V)} \leq e^{Ct} \|f_0\|_{L^2(\mathbb{R}^d \times V)}, \tag{5.49}$$

$$\frac{1}{\varepsilon^2} \int_0^T \int_{\mathbb{R}^d \times V \times V} |f_\varepsilon(t, x, \xi) - f_\varepsilon(t, x, \xi')|^2 dx d\xi d\xi' dt \leq 2(1 + e^{2CT})\|f_0\|_{L^2(\mathbb{R}^d \times V)},$$
$$\tag{5.50}$$

with

$$C(T) = \sup_{0 \le t \le T, x \in \mathbb{R}^d} \int_V \sup_{\xi' \in V} \frac{K_r^2}{K_s}(c; \xi, \xi') d\xi.$$

We recall that the dependency on (t, x) is only through c in the above formula.

Proof. We multiply (5.45) by f_ε and integrate in x and ξ. We obtain

$$\frac{1}{2}\frac{d}{dt}\int_{\mathbb{R}^d \times V} f_\varepsilon(t)^2$$
$$= -\frac{1}{2\varepsilon^2}\int_{\mathbb{R}^d \times V \times V} K_s(c; \xi, \xi') |f_\varepsilon(t, x, \xi) - f_\varepsilon(t, x, \xi')|^2 dx d\xi d\xi'$$
$$+ \frac{1}{\varepsilon}\int_{\mathbb{R}^d \times V \times V} K_{r,\varepsilon}(c; \xi, \xi') \left(f_\varepsilon(t, x, \xi) - f_\varepsilon(t, x, \xi')\right) f_\varepsilon(t, x, \xi') dx d\xi d\xi'$$
$$\le -\frac{1}{2\varepsilon^2}\int_{\mathbb{R}^d \times V \times V} k_m |f_\varepsilon(t, x, \xi) - f_\varepsilon(t, x, \xi')|^2 dx d\xi d\xi'$$
$$+ \frac{1}{4\varepsilon^2}\int_{\mathbb{R}^d \times V \times V} k_m \left(f_\varepsilon(t, x, \xi) - f_\varepsilon((t, x, \xi'))\right)$$
$$+ \int_{\mathbb{R}^d \times V \times V} \frac{K_{r,\varepsilon}(c; \xi, \xi')^2}{k_m} f_\varepsilon(t, x, \xi')^2 dx d\xi d\xi'$$
$$\le -\frac{1}{4\varepsilon^2}\int_{\mathbb{R}^d \times V \times V} k_m |f_\varepsilon(t, x, \xi) - f_\varepsilon(t, x, \xi')|^2 dx d\xi d\xi'$$
$$+ C(T)\int_{\mathbb{R}^d \times V} f_\varepsilon(t, x, \xi')^2 dx d\xi'.$$

Therefore, we may directly apply Gronwall's lemma and obtain the first inequality (5.49). Then, integrating in time the above inequality, and using (5.49), we obtain (5.50). □

Notice that this L^2 bound has consequences. The first one is that $n_\varepsilon = \int_V f_\varepsilon d\xi$ is also bounded in L^2. Therefore in dimensions 2 and 3 this implies that the chemoattactant c_ε is bounded in L^∞. It is proved in [54] that, by a Gronwall argument this implies

Corollary 5.1. *In dimension $d = 2$ or 3, with the assumption of Proposition 5.1, the turning kernels (5.40) and $f_0 \in L^2(\mathbb{R}^d \times V)$, there is a time $T^* > 0$ such that the solution to (5.49) is uniformly (in ε and $0 \le t \le T^*$) bounded in $L^2(\mathbb{R}^d \times V)$ and $c_\varepsilon(t, x)$ is uniformly bounded in $L^\infty(\mathbb{R}^d)$. Moreover the estimates (5.49) and (5.50) hold up to this time T^*.*

With this in mind we can proceed and study the limit as ε vanishes. At this point we would like to recall that this limiting process leads to the Keller–Segel model that blows-up in finite time. Therefore the result Corollary 5.1 can indeed hold only on a finite time interval.

5.7.3   Limit as ε vanishes

We first draw some conclusions from the a priori bounds in Proposition 5.1. The proof of Theorem 5.8 relies on the analysis of the consequences of Proposition 5.1 and on the relation

$$\frac{\partial n_\varepsilon}{\partial t} + \operatorname{div} J_\varepsilon = 0, \qquad 0 \le t \le T^*, \ x \in \mathbb{R}^d, \tag{5.51}$$

with

$$n_\varepsilon(t,x) = \int_V f_\varepsilon(t,x,\xi)d\xi, \qquad J_\varepsilon = \frac{1}{\varepsilon}\int_V \xi\, f_\varepsilon(t,x,\xi)d\xi. \qquad (5.52)$$

This is obtained integrating (5.49) in ξ.

Lemma 5.3. *Recalling that $\int_V \xi\, d\xi = 0$ and $|V| = 1$, for some $n \in L^\infty\big((0,T^*);$ $L^2(\mathbb{R}^d)\big)$, $J \in L^2\big((0,T^*) \times \mathbb{R}^d\big)$, we have after extraction of a subsequence $\varepsilon_n \to 0$,*

$$n_\varepsilon(t,x,\xi) \rightharpoonup n(t,x), \quad \text{weakly in } L^2\big((0,T^*) \times \mathbb{R}^d\big),$$

$$f_\varepsilon(t,x,\xi) \rightharpoonup n(t,x)\,\mathbf{1}_{\{\xi \in V\}}, \quad \text{weakly in } L^2\big((0,T^*) \times \mathbb{R}^d \times V\big),$$

$$J_\varepsilon(t,x,\xi) \rightharpoonup J(t,x), \quad \text{weakly in } L^2\big((0,T^*) \times \mathbb{R}^d\big).$$

Proof. From the $L^\infty\big((0,T^*); L^2(\mathbb{R}^d \times V)\big)$ bound on f_ε, we deduce also that n_ε is bounded in $L^\infty\big((0,T^*); L^2(\mathbb{R}^d)\big)$, and thus admits a subsequence that converges weakly to some $n \in L^\infty\big((0,T^*); L^2(\mathbb{R}^d)\big)$ as indicated in the first result.

Next, we use the estimate (5.50) and thus, thanks to the Cauchy–Schwarz inequality with $|V| = 1$, we have

$$\int_0^{T^*}\int_{\mathbb{R}^d \times V} |f_\varepsilon(t,x,\xi) - n_\varepsilon(t,x)|^2 dxd\xi dt$$
$$= \int_0^{T^*}\int_{\mathbb{R}^d \times V} |\int_V [f_\varepsilon(t,x,\xi) - f_\varepsilon(t,x,\xi')]d\xi'|^2 dxd\xi dt$$
$$\leq \int_0^{T^*}\int_{\mathbb{R}^d \times V \times V} |f_\varepsilon(t,x,\xi) - f_\varepsilon(t,x,\xi')|^2 dxd\xi d\xi' dt$$
$$\leq C\varepsilon^2.$$

This proves that, after further extraction, the weak limit of f_ε is the same as that of n_ε, the second statement.

With the same type of argument, since $\int_V \xi\, d\xi = 0$, we have

$$\int_0^{T^*}\int_{\mathbb{R}^d} |J_\varepsilon(t,x)|^2 dxdt = \int_0^{T^*}\int_{\mathbb{R}^d} |\int_{V \times V} \frac{\xi}{\varepsilon}[f_\varepsilon(t,x,\xi) - f_\varepsilon(t,x,\xi')]d\xi d\xi'|^2 dxdt$$
$$\leq \frac{C}{\varepsilon^2}\int_0^{T^*}\int_{\mathbb{R}^d \times V \times V} |f_\varepsilon(t,x,\xi) - f_\varepsilon(t,x,\xi')|^2 dxd\xi d\xi' dt$$
$$\leq C$$

and thus $J_\varepsilon(t,x)$ is bounded in $L^2\big((0,T^*) \times \mathbb{R}^d\big)$ and thus admits a weak limit also. $\qquad\square$

Thanks to this lemma, we may pass to the limit in (5.51), and obtain

$$\frac{\partial n}{\partial t} + \text{div}J = 0, \qquad 0 \leq t \leq T^*,\ x \in \mathbb{R}^d. \qquad (5.53)$$

In order to conclude the proof of Theorem 5.8, it remains to identify the 'flux' $J(t,x)$. To do so we write the equation (5.49) as

$$\frac{1}{\varepsilon}\int_V K_s(c_\varepsilon;\xi,\xi')\,[f_\varepsilon(t,x,\xi') - f_\varepsilon(t,x,\xi)]d\xi' = \varepsilon\frac{\partial}{\partial t}f_\varepsilon + \xi \cdot \nabla_x f_\varepsilon$$
$$+ \int_V \big[K_{r,\varepsilon}(c_\varepsilon;\xi',\xi)f_\varepsilon(t,x,\xi) - K_{r,\varepsilon}(c_\varepsilon;\xi,\xi')f_\varepsilon(t,x,\xi')\big]d\xi'.$$

We integrate against the weight $\xi\, d\xi$ and obtain (using the specific form of $K_s = k_+(c_\varepsilon) + k_-(c_\varepsilon)$ and $\int_V \xi\, d\xi = 0$),

$$-[k_+(c_\varepsilon) + k_-(c_\varepsilon)]J_\varepsilon(t,x) = \varepsilon\frac{\partial n_\varepsilon}{\partial t} + \text{div} \int_V \xi \otimes \xi f_\varepsilon(t,x,\xi)d\xi + J_{a,\varepsilon}(t,x),$$

with

$$J_{a,\varepsilon}(t,x) = \int_{V\times V} \xi\big[K_{r,\varepsilon}(c_\varepsilon;\xi',\xi)f_\varepsilon(t,x,\xi) - K_{r,\varepsilon}(c_\varepsilon;\xi,\xi')f_\varepsilon(t,x,\xi')\big]d\xi'd\xi.$$

Because of the weak limits in Lemma 5.3, and because c_ε converges strongly [3], we may pass to the limit in the distribution sense and obtain

$$-[k_+(c) + k_-(c)]J(t,x) = \bar{D}\cdot\nabla n + J_a,$$

with $\bar{D}$ the matrix of entries $\int_V \xi \otimes \xi d\xi$ which is enough to deduce the diffusion matrix $D = \bar{D}/\big(k_+(c) + k_-(c)\big)$ in Theorem 5.8. It remains to compute J_a,

$$J_a(t,x) = n(t,x)\lim_{\varepsilon\to 0} \int_{V\times V} \xi\big[K_{r,\varepsilon}(c;\xi',\xi) - K_{r,\varepsilon}(c;\xi,\xi')\big]d\xi'd\xi.$$

And from (5.40) and (5.48), we have

$$\int_{V\times V} \xi\big[K_{r,\varepsilon}(c;\xi',\xi) - K_{r,\varepsilon}(c;\xi,\xi')\big]d\xi'd\xi$$
$$\approx \int_{V\times V} \xi\big[k'_-(c)(-\xi + \xi') + k'_+(c)(\xi' - \xi)\big].\nabla c\, d\xi'd\xi.$$

This leads to the following components of the vector J_a,

$$J_{a,i}(t,x) = -n(t,x)\frac{\partial c}{\partial x_j}\big[k'_-(c) + k'_+(c)\big]\int_V \xi_i\, \xi_j\, d\xi.$$

Altogether, we obtain the coefficients announced in Theorem 5.8. $\qquad\square$

5.8 Hydrodynamic limit

The mere scattering model (5.38), (5.40), (5.39) does not allow us to derive a hydrodynamic limit. Indeed local interactions (in the spirit of the binary collision operator in Boltzmann equation, [53]) are needed for such a derivation. Local interactions between cells in the kinetic model are not presented in the literature but several hints that they might exist are based on biochemical investigations. This pushed [103] to postulate a variant of the scattering equation where, additionally to the long range interaction due to chemoattraction, some local operator is introduced, thus arriving at a BGK type model (the case of a Vlasov equation is treated in [122]). Those models have the advantage of avoiding the physical description of

---

[3]Here there are technical difficulties; namely some uniform regularity of $K_{r,\varepsilon}$ in c has to be assumed, and the regularizing effect on c_ε in space through the Laplace equation is not enough and time has to be treated also, see [54].

the local interactions since they only require us to know the equilibrium state (see the comment at the beginning of Section 5.7)

$$\frac{n}{\vartheta^{d/2}(n)} F\left(\frac{\xi - u}{\vartheta^{1/2}(n)}\right).$$

Here $F : \mathbb{R} \to [0, \infty)$ is a given smooth function, $\vartheta : [0, \infty) \to [0, \infty)$ is a power-like function and we use now $V = \mathbb{R}^d$ because Galilean invariance is fundamental in this approach. In order to fullfill basic conservation laws (number of cells and momentum) one assumes that

$$\int_{\mathbb{R}^d} F(\eta)\, d\eta = 1, \quad \text{i.e.} \quad \int_{\mathbb{R}^d} \frac{n}{\vartheta^{d/2}(n)} F\left(\frac{\xi - u}{\vartheta^{1/2}(n)}\right) d\xi = n,$$

$$\int_{\mathbb{R}^d} \eta F(\eta)\, d\eta = 0, \quad \text{i.e.} \quad \int_{\mathbb{R}^d} \xi \frac{n}{\vartheta^{d/2}(n)} F\left(\frac{\xi - u}{\vartheta^{1/2}(n)}\right) d\xi = nu.$$

We also need notation for the macroscopic quantities defined as

$$n_\varepsilon = \int_{\mathbb{R}^d} f_\varepsilon(t, x, \xi)\, d\xi, \qquad n_\varepsilon u_\varepsilon = \int_{\mathbb{R}^d} \xi f_\varepsilon(t, x, \xi)\, d\xi.$$

The quantity $u_\varepsilon(t, x)$ is therefore the average (bulk) velocity of the cells at time t and point x while ξ is their microscopic density.

With this notation, the model proposed in [103] reads

$$\frac{\partial}{\partial t} f_\varepsilon(t, x, \xi) + \xi \cdot \nabla_x f_\varepsilon + \int_{\mathbb{R}^d} K_1(\xi, \xi', c) \cdot \nabla c\, f(\xi')\, d\xi'$$
$$= \int_{\mathbb{R}^d} K_1(\xi', \xi, c) \cdot \nabla c\, d\xi'\, f(\xi). - \frac{1}{\varepsilon}\left[f_\varepsilon - \frac{n}{\vartheta^{d/2}(n)} F\left(\frac{\xi - u}{\vartheta^{1/2}(n)}\right)\right].$$
$$(5.54)$$

The scattering kernel has been modified to take into account only the antisymmetric part of K in (5.38) because the main collision operator (zeroth order part of K) has been replaced by the mere relaxation to the equilibrium. Doing that, one can notice an important flaw in the model, that nonnegativity of f_ε is lost because the turning crosssection $K_1(v, v', c) \cdot \nabla c$ has no sign. This disqualifies the model but the mere relaxation is already an artefact. However it keeps the fundamental idea and motivation behind this model; (i) scattering only describes interactions with an external medium (or long range interactions through the chemoattractant) and thus is fundamentally a linear operator for a given function $c(t, x)$, (ii) but here we wish to model local self interactions of cells and this requires a nonlinear local operator (namely the relaxation to the equilibrium state). Additionally, we have used a hyperbolic scale for the rescaling of (5.38) rather than a parabolic scale as was done in Section 5.7.

In the so-called hydrodynamic limit, i.e., $\varepsilon \to 0$ in (5.54), we then obtain the following model for the cell movements

$$
\begin{cases}
\dfrac{\partial n}{\partial t} + \operatorname{div}(n\,u) = 0, \\[2mm]
\dfrac{\partial(nu)}{\partial t} + \operatorname{div}\big(n\,u \otimes u + n\vartheta(n)p\big) = n\,\vartheta^{(d+1)/2}(n)\,\chi(n,u,c)\nabla c,
\end{cases}
\tag{5.55}
$$

still coupled with the concentration equation $-\Delta c = n$ (or whatever has been supposed for the production of c). The matrix χ is given by

$$
\chi(n,u,S) = \int_{V \times V} (v - v') \otimes K_1(u + \vartheta^{1/2}v, u + \vartheta^{1/2}v', S)F(v')\,dv'\,dv.
$$

This type of hyperbolic system is not new. It is reminiscent of fluid dynamics (for a compressible gas with gravitational forces, these are models used in astrophysics). In the context of biology, it has been proposed to describe the initiation of angiogenesis in [113, 209]. Indeed solutions to this system exhibit network structures very close to the networks formed by endothelial cells. It is therefore compatible with the idea that blood vessel formation is related with local interactions of cells as one can see in recent experiments. This is not the case of the Keller–Segel model that describes only long range interactions and does not give network structures. Numerical simulations of these networks are presented in Figure 5.7 and taken from from [104] where an accurate algorithm has been elaborated (see also [113, 209, 103]).

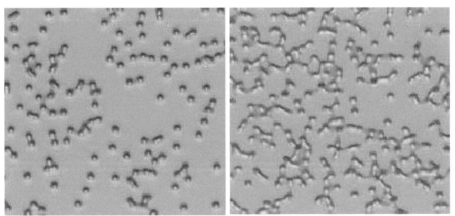

Figure 5.7: NETWORK FORMATION IN THE HYPERBOLIC SYSTEM OF CHEMOTAXIS (5.55) FOR TWO DIFFERENT INITIAL DENSITIES (THE FIGURE ON THE RIGHT CORRESPONDS TO TWICE THE DENSITY IN THE FIGURE ON THE LEFT). FIGURE TAKEN FROM [104].

Chapter 6

General mathematical tools

In this chapter we present several general mathematical tools that have been used throughout these notes. For several topics, we try to go further than the mere results that are actually needed. This is aimed at presenting also the context of mathematical research around the formalisms used in the models we have presented.

6.1 Transport equations and the method of characteristics

6.1.1 Transport equation (strong form)

We first consider the transport equation written in the strong form (in contrast to the conservative, divergence or weak form treated in Section 6.1.2 below)

$$\begin{cases} \frac{\partial}{\partial t} u(t,x) + b(t,x) \cdot \nabla u(t,x) = 0, & t \in \mathbb{R}, \ x \in \mathbb{R}^d, \\ u(t=0) = u^0 \quad \text{given.} \end{cases} \tag{6.1}$$

We always assume here that the vector field $b \in C^1(\mathbb{R}^{1+d}; \mathbb{R}^d)$ satisfies the Cauchy–Lipschitz conditions. Namely, for all $T > 0$, $R > 0$ there are two constants $M_1(T,R)$, $M_2(T)$ such that

$$\begin{cases} |b(t,x) - b(t,y)| \leq M_1(T,R)|x-y|, & \text{for} \quad |t| \leq T, \ |x|, \ |y| \leq R, \\ |b(t,x)| \leq M_2(T)(1+|x|), & \text{for} \quad |t| \leq T, \ x \in \mathbb{R}^d. \end{cases} \tag{6.2}$$

This assumption allows us to define the so-called *characteristics* associated with equation (6.1),

$$\begin{cases} \dot{X}(t;y) = b\big(t, X(t;y)\big), \\[2mm] X(0;y) = y \in \mathbb{R}^d. \end{cases} \tag{6.3}$$

They are defined for all times $t \in \mathbb{R}$ and the mapping $y \mapsto X(t;y)$ is a C^1 diffeomorphism on $\mathbb{R}^d$. These are deeply related to the existence of solutions to the transport equation (6.1).

Theorem 6.1 (Smooth solutions). *Under assumptions (6.2), $b \in C^1(\mathbb{R} \times \mathbb{R}^d)$ and $u^0 \in C^1(\mathbb{R}^d)$, there is a unique $C^1(\mathbb{R} \times \mathbb{R}^d)$ solution to (6.1) and it is constant along the characteristics i.e.*

$$u\big(t, X(t;y)\big) = u^0(y), \quad \forall t \in \mathbb{R}, \ \forall y \in \mathbb{R}^d. \tag{6.4}$$

Proof. We have

$$\begin{aligned} \tfrac{d}{dt} u\big(t, X(t;y)\big) &= \tfrac{\partial}{\partial t} u\big(t, X(t;y)\big) + \dot{X}(t;y) \cdot \nabla u\big(t, X(t;y)\big) \\ &= \tfrac{\partial}{\partial t} u\big(t, X(t;y)\big) + b\big(t, X(t;y)\big) \cdot \nabla u\big(t, X(t;y)\big). \end{aligned}$$

Therefore, the transport equation holds true if and only if $u\big(t, X(t;y)\big)$ is independent of time.

The C^1 regularity follows from the differentiability of the flow, because the $d \times d$ matrix $D_y X(t;y)$ satisfies the differential equation

$$\tfrac{d}{dt} D_y X(t;y) = D_x b(t, X(t;y)) D_y X(t;y) = 0 \qquad D_y X(t=0;y) = I_d. \qquad \square$$

We can deduce some general properties of the solutions to transport equation in the strong form. These properties are typical of the *hyperbolic* nature of transport equations. They are obvious consequences of the representation formula in Theorem 6.1 and we do not prove them.

Proposition 6.1. *The solution of (6.1) satisfies:*

(i) *(Finite speed of propagation) The value $u(t,x)$ only depends on the values of $u^0(y)$ for $|y - x| \le |t| \, \|b\|_{L^\infty}$.*

(ii) *(Propagation of singularities) The solution $u(t,\cdot)$ has the same regularity as u^0 (and not more).*

(iii) *(Maximum principle) For all $t \in \mathbb{R}$, $x \in \mathbb{R}^d$, we have*

$$\inf_{y \in \mathbb{R}^d} u^0(y) \le u(t,x) \le \sup_{y \in \mathbb{R}^d} u^0(y).$$

Theorem 6.2 (Weak solutions). *Under assumptions (6.2), $b \in C^1(\mathbb{R} \times \mathbb{R}^d)$ and $u^0 \in L^\infty(\mathbb{R}^d)$, there is a unique solution in the distribution sense $u \in L^\infty(\mathbb{R}^{d+1}) \cap C\big(\mathbb{R}^+; L^1_{\mathrm{loc}}(\mathbb{R}^d)\big)$ to (6.1) and it is constant along the characteristics i.e. (6.4) holds true.*

We recall

Definition 6.1. A solution $u(t,x)$ to (6.1) in distribution sense is defined by integration by parts against smooth functions, i.e., by the equality

$$-\int_0^\infty \int_{\mathbb{R}^d} u(t,x)\Big[\frac{\partial}{\partial t}\varphi(t,x) + \text{div}\,\big(b(t,x)\varphi(t,x)\big)\Big]dxdt = \int_{\mathbb{R}^d} \varphi(t=0,x)\,u^0(x)dx,$$
(6.5)

for all test functions $\varphi \in \mathcal{D}(\mathbb{R}\times\mathbb{R}^d)$ (and similarly for negative times).

Proof. Existence. We define a regularized sequence of initial data $u_n^0 = u^0 \star \rho_n \in C^\infty(\mathbb{R}^d)$, for some regularizing kernel ρ_n. We have $u_n^0 \to u^0$ a.e. and in $L^1_{\text{loc}}(\mathbb{R}^d)$ and $|u_n^0| \le \|u^0\|_{L^\infty}$. We denote by $u_n(t,x)$ the sequence of C^1 solutions to (6.1).

On the one hand, since the characteristics define a diffeomorphism, one has for all $T > 0$,

$$\|u_n(t,x) - u_m(t,x)\|_{C((-T,T);L^1_{\text{loc}}(\mathbb{R}^d))} \to 0.$$

Therefore $u_n(t,x)$ is a Cauchy sequence in $C\big(-T,T;L^1_{\text{loc}}(\mathbb{R}^d)\big)$ and thus it converges in this space to a function $u \in C\big(\mathbb{R};L^1_{\text{loc}}(\mathbb{R}^d)\big)$. On the other hand, the corresponding solution $u_n(t,x)$ given by Theorem 6.1 is also a solution in distribution sense (thanks to usual integration by parts). Therefore, we have for all test functions $\varphi \in \mathcal{D}(\mathbb{R}\times\mathbb{R}^d)$,

$$-\int_0^\infty \int_{\mathbb{R}^d} u_n(t,x)\Big[\tfrac{\partial}{\partial t}\varphi(t,x) + \text{div}\,\big(b(t,x)\varphi(t,x)\big)\Big]dxdt = \int_{\mathbb{R}^d} \varphi(t=0,x)\,u_n^0(x)dx,$$

and passing to the limit thanks to the Lebesgue theorem, we find (6.5).

Uniqueness. Because of time continuity in $L^1_{\text{loc}}(\mathbb{R}^d)$, we can first establish the following variant of (6.5),

$$\begin{aligned}-\int_0^T \int_{\mathbb{R}^d} u(t,x)[\tfrac{\partial}{\partial t}\varphi(t,x) &+\text{div}\,(b(t,x)\varphi(t,x))]dxdt \\ &= \int_{\mathbb{R}^d} \varphi(0,x)\,u^0(x)dx - \int_{\mathbb{R}^d}\varphi(T,x)\,u(T,x)dx,\end{aligned}$$
(6.6)

for all $T > 0$ and all test functions $\varphi \in C^1(\mathbb{R}\times\mathbb{R}^d)$ with compact support in space (and similarly for negative times). For the difference $u = u_1 - u_2$ of two solutions with the same initial data, we therefore obtain

$$-\int_0^T \int_{\mathbb{R}^d} u(t,x)\Big[\tfrac{\partial}{\partial t}\varphi(t,x) + \text{div}\,\big(b(t,x)\varphi(t,x)\big)\Big]dxdt = -\int_{\mathbb{R}^d}\varphi(T,x)\,u(T,x)dx.$$

But, as we see in the next section, one can solve the 'dual' or 'backward' equation

$$\tfrac{\partial}{\partial t}\varphi(t,x) + \text{div}\,\big(b(t,x)\varphi(t,x)\big) = 0, \qquad \varphi(T,x) \in C^1_{\text{comp}}(\mathbb{R}^d) \text{ given.}$$

Its solution is C^1 and has compact support in space, therefore we can use it in (6.6) and finally obtain

$$0 = -\int_{\mathbb{R}^d}\varphi(T,x)\,u(T,x)dx,$$

and this holds for all test functions $\varphi(T,x) \in C^1_{\text{comp}}(\mathbb{R}^d)$. This proves the result. $\square$

6.1.2   Transport equation (conservative form)

We can also consider the conservative form of the transport equation

$$\begin{cases} \frac{\partial}{\partial t} u(t,x) + \operatorname{div}\big(b(t,x)u(t,x)\big) = 0, & t \in \mathbb{R},\ x \in \mathbb{R}^d, \\[2mm] u(t=0) = u^0 & \text{given.} \end{cases} \tag{6.7}$$

In order to give a representation formula, we introduce the notation

$$J(t;y) = \exp\Big(\int_0^t \operatorname{div} b(s, X(s;y))\,ds\Big), \qquad t \in \mathbb{R},\ y \in \mathbb{R}^d,$$

which is nothing but the solution to the differential equation

$$\dot{J}(t;y) = \operatorname{div} b\big(t, X(t;y)\big)\, J(t;y), \qquad J(0,y) = 1.$$

In other words, thanks to the Liouville theorem,

$$J(t;y) = \operatorname{Jac}\Big(\frac{\partial X(t;y)}{\partial y}\Big).$$

Theorem 6.3. *Under assumptions (6.2), b and $\operatorname{div} b \in C^1$ and $u^0 \in C^1(\mathbb{R}^d)$, there is a unique C^1 solution to (6.7) and it satisfies the representation formula*

$$u\big(t, X(t;y)\big)\, J(t;y) = u^0(y), \qquad \forall t \in \mathbb{R},\ \forall y \in \mathbb{R}^d. \tag{6.8}$$

Proof. We write the conservative transport equation in the form

$$\tfrac{\partial}{\partial t} u(t,x) + b(t,x) \cdot \nabla u(t,x) + \operatorname{div} b(t,x)\, u(t,x) = 0,$$

and thus it is equivalent to writing

$$\tfrac{d}{dt} u\big(t, X(t;y)\big) + \operatorname{div} b\big(t, X(t;y)\big)\, u\big(t, X(t;y)\big) = 0,$$

which is also equivalent to writing

$$\tfrac{d}{dt}\big[u\big(t, X(t;y)\big)\, J(t;y)\big] = 0.$$

$\square$

We also deduce from this representation the following abstract. properties

Proposition 6.2. *The solution of (6.7) satisfies the properties:*

(i) *(Finite speed of propagation) The value $u(t,x)$ only depends on the values of $u^0(y)$ for $|y - x| \le |t|\, \|b\|_{L^\infty}$.*

(ii) *(Propagation of singularities) The solution $u(t,\cdot)$ has the same regularity as u^0 (and not more).*

And for $u^0 \in L^1(\mathbb{R}^d)$,

(iii) *(Mass conservation)*

$$\int_{x \in \mathbb{R}^d} u(t, x) = \int_{y \in \mathbb{R}^d} u^0(y).$$

(iv) *(L^1 stability)*

$$\int_{x \in \mathbb{R}^d} |u(t, x)| = \int_{y \in \mathbb{R}^d} |u^0(y)|.$$

Statements (i) and (ii) are again obvious consequences of the representation formula (6.8). As for the statements (iii) and (iv), they are consequences, after a change of variables, of the fact that J is the Jacobian of the diffeomorphism $y \mapsto X(t; y)$. But we can also see it directly when u^0 has a bounded support. Then $u(t, x)$ also has bounded support and we can write, thanks to the Lebesgue theorem and Green's formula,

$$\frac{d}{dt} \int_{x \in \mathbb{R}^d} u(t, x) = \int_{x \in \mathbb{R}^d} \frac{\partial}{\partial t} u(t, x) = \int_{x \in \mathbb{R}^d} \mathrm{div}\big(b\, u(t, x)\big) = 0.$$

As for the absolute value, we can see it in the same way but we first need to notice

Proposition 6.3. *Under the assumptions of Theorem 6.3, we have in distribution sense*

$$\begin{cases} \frac{\partial}{\partial t} |u(t, x)| + \mathrm{div}\,\big(b(t, x)|u(t, x)|\big) = 0, & t \in \mathbb{R}, \ x \in \mathbb{R}^d, \\ |u(t = 0)| = |u^0|. \end{cases} \tag{6.9}$$

In other words, for all test functions $\varphi \in \mathcal{D}(\mathbb{R}^{1+d})$, we have

$$-\int_0^\infty \int_{x \in \mathbb{R}^d} |u(t, x)| \left(\frac{\partial}{\partial t}\varphi + b \cdot \nabla\varphi\right) = \int_{x \in \mathbb{R}^d} |u^0(x)| \, \varphi(t = 0, x) dx.$$

Proof. We choose the C^1 function

$$S_\delta(u) = \begin{cases} \frac{u^2}{2\delta} & \text{for } |u| \leq \delta, \\ |u| - \frac{\delta}{2} & \text{for } |u| \geq \delta. \end{cases}$$

Using the chain rule, we have

$$\frac{\partial}{\partial t} S_\delta(u) + \mathrm{div}\,\big(b(t, x)S_\delta(u)\big) + \mathrm{div}\,\big(b(t, x)\big) \left[uS'_\delta(u) - S_\delta(u)\right] = 0.$$

As $\delta \to 0$, we have $S_\delta(u) \to |u|$ strongly in L^1_{loc} and, since $|uS'_\delta(u) - S_\delta(u)| \leq \delta/2$, we obtain the result. $\square$

Proposition 6.4. *Under the assumptions* (6.2), *b and* $\operatorname{div} b \in C^1$, *consider the measure*

$$u(t,x) = \delta\big(x - X(t;y)\big), \qquad X(0;y) = y \in \mathbb{R}^d.$$

It is a solution in the distribution sense to equation (6.7) *if and only if* $X(t;y)$ *is a solution to the differential equation* (6.3).

Proof. The measure $u(t,x)$ is a distributional solution means that for all test functions $\varphi(t,x)$, the identity holds:

$$-\int_0^\infty \int_{\mathbb{R}^d} u(t,x)\big[\tfrac{\partial}{\partial t}\varphi + b(t,x)\cdot\nabla\varphi\big]dx\,dt = \int_{\mathbb{R}^d} u^0(x)\varphi(x)dx.$$

With the Dirac solution, we obtain

$$-\int_0^\infty \tfrac{\partial}{\partial t}\varphi\big(t,X(t;y)\big) + b\big(t,X(t;y)\big)\cdot\nabla\varphi\big(t,X(t;y)\big)\,dt = \varphi(y).$$

Using the method of characteristics in 6.1.1, we deduce

$$-\int_0^\infty \tfrac{d}{dt}\varphi\big(t,X(t;y)\big) + \big[b\big(t,X(t;y)\big) - \dot{X}(t;y)\big]\cdot\nabla\varphi\big(t,X(t;y)\big)\,dt = \varphi(y).$$

And we conclude that

$$\int_0^\infty \big[b\big(t,X(t;y)\big) - \dot{X}(t;y)\big]\cdot\nabla\varphi\big(t,X(t;y)\big)\,dt = 0$$

for all test functions φ, which leads to the conclusion of Proposition 6.4. □

6.2   Transport equation (the DiPerna–Lions theory)

It is possible to improve the above results and optimize the assumptions on the velocity field b while keeping existence and uniqueness. This theory is due to DiPerna and Lions [84], see also [158]. We present for instance the case of the equation under its strong form but many variants are possible, including for instance first order terms or source terms. Also, we restrict ourselves to the theory $\nabla b \in L^1$ theory, but we mention the recent extension to $\nabla b \in BV$ in [7].

Theorem 6.4. *Let $b \in L^\infty_{\mathrm{loc}}(\mathbb{R}^d)$, $\nabla b \in L^1(\mathbb{R}^d)$ and $\operatorname{div} b \in L^\infty(\mathbb{R}^d)$; then there exists a unique solution $u \in L^\infty(\mathbb{R} \times \mathbb{R}^d)$ to the equation*

$$\begin{cases} \tfrac{\partial}{\partial t}u(t,x) + b(t,x)\cdot\nabla u(t,x) = 0, & t \in \mathbb{R},\ x \in \mathbb{R}^d, \\[2mm] u(t=0) = u^0 \in L^\infty(\mathbb{R}^d). \end{cases} \tag{6.10}$$

The notion of solution for the form (6.10) is not completely obvious and we have in mind to rewrite the equation as

$$\frac{\partial}{\partial t}\varphi(t,x) + \operatorname{div}\big(b(t,x)\varphi(t,x)\big) - (\operatorname{div} b)u(t,x) = 0.$$

6.2.1 A regularization Lemma

Before proving this theorem, we need a commutation lemma which is the heart of the proof.

Lemma 6.1. *Let* $\varphi \in L^{\infty}\left((-T,T) \times \mathbb{R}^d\right)$, $\forall T > 0$, *a solution of the transport equation*

$$\frac{\partial}{\partial t}\varphi(t,x) + \operatorname{div}\left(b(t,x)\varphi(t,x)\right) = 0, \quad t \in \mathbb{R}, \ x \in \mathbb{R}^d. \tag{6.11}$$

Then, for any regularizing kernel $\rho_\varepsilon(x) = \frac{1}{\varepsilon^d}\rho(\frac{x}{\varepsilon})$ *with* $\rho \in \mathcal{D}(\mathbb{R}^d)$, $\rho \geq 0$, $\int \rho = 1$, *the convolution* $\varphi_\varepsilon = \varphi \star \rho_\varepsilon$ *satisfies*

$$\frac{\partial}{\partial t}\varphi_\varepsilon(t,x) + \operatorname{div}\left(b(t,x)\varphi_\varepsilon(t,x)\right) = r_\varepsilon(t,x) \to 0 \quad in \quad L^{\infty}\left((-T,T) \times \mathbb{R}^d\right).$$

Proof. We write

$$\begin{aligned}
r_\varepsilon(t,x) &= \operatorname{div}(b\varphi_\varepsilon) - \operatorname{div}(b\varphi) \star \rho_\varepsilon \\
&= (\operatorname{div}b)\varphi_\varepsilon + b \cdot \nabla\varphi_\varepsilon - (b\varphi) \star \nabla\rho_\varepsilon \\
&= (\operatorname{div}b)\varphi_\varepsilon + \int_{\mathbb{R}^d}[b(x) - b(x-y)]\varphi(x-y)\nabla\rho_\varepsilon(y)\,dy \\
&= (\operatorname{div}b)\varphi_\varepsilon + \int_0^1\int_{\mathbb{R}^d}\nabla b(x-sy) \cdot y\nabla\rho_\varepsilon(y)\varphi(x-y)\,dy \\
&= R_\varepsilon + S_\varepsilon,
\end{aligned}$$

with

$$R_\varepsilon(t,x) = (\operatorname{div}b)\varphi_\varepsilon + \nabla b(x) \cdot \left(y \otimes \nabla\rho_\varepsilon(y)\right) \star \varphi \to 0 \quad \in L^{\infty}\left((-T,T); L^1(\mathbb{R}^d)\right),$$

because $y \otimes \nabla\rho_\varepsilon(y)$ is a regularizing kernel that satisfies $\int y_i\nabla_{y_j}\rho(y)\,dy = 0$ for $i \neq j$ and $\int y_i\nabla_{y_i}\rho(y)\,dy = -1$.

On the other hand

$$S_\varepsilon(t,x) = \int_0^1\int_{\mathbb{R}^d}[\nabla b(x-sy) - \nabla b(x)] \cdot y\nabla\rho_\varepsilon(y)\varphi(x-y)\,dy,$$

and thus, on an interval $(-T,T)$ we have

$$\begin{aligned}
|S_\varepsilon(t,x)| &\leq \|\varphi\|_{L^\infty}\int_0^1\int_{\mathbb{R}^d}|\nabla b(x-sy) - \nabla b(x)|\,|y\nabla\rho_\varepsilon(y)|\,dy \\
&\leq \|\varphi\|_{L^\infty}\int_{\mathbb{R}^d}\omega(\nabla b; |y|)\,|y\nabla\rho_\varepsilon(y)|\,dy \\
&\longrightarrow 0 \quad in \quad L^\infty(-T,T) \quad as \ \varepsilon \to 0,
\end{aligned}$$

where the L^1 modulus of continuity is defined by $\omega(\nabla b; h) = \sup_{|y|\leq h}|\nabla b(x+y) - \nabla b(x)|$ and satisfies $\omega(\nabla b; 0) = 0$; this is the place where we explicitly use $\nabla b \in L^1$ and not a mere bounded measure. $\qquad\square$

Remark 6.1. For time dependent velocity fields b, the corresponding assumption and conclusions are

$$\nabla b \in L^1\left((-T,T) \times \mathbb{R}^d\right), \qquad r_\varepsilon \to 0 \quad \text{in} \ \ L^1((-T,T) \times \mathbb{R}^d).$$

Indeed, in the above argument we just have to control the L^1 modulus of continuity in the space of ∇b.

6.2.2   Proof of Theorem 6.4

Existence. We argue by regularization. Consider b_ε a family of smooth vector fields with compact support such that b_ε is locally bounded, $\mathrm{div} b_\varepsilon$ bounded in L^∞, $b_\varepsilon \to b$ a.e. and $\nabla b_\varepsilon \to \nabla b$ in L^1. Then, using Section 6.1.1, there exists a unique solution u_ε to the transport equation in strong form for this velocity field b_ε. From the method of characteristics we have $|u_\varepsilon| \le \|u^0\|_{L^\infty}$, and thus we can extract a subsequence which converges in $L^\infty w - \star$ to a function $u(t,x)$. Then we may pass to the limit in the definition of weak solutions (6.5) and existence is proved.

Uniqueness. For uniqueness we consider a solution in distribution sense with vanishing initial data. Therefore, for all test function $\psi \in \mathcal{D}(\mathbb{R} \times \mathbb{R}^d)$, we have

$$\int_0^\infty u(t,x) \left[\frac{\partial}{\partial t}\psi(t,x) + \mathrm{div}\left(b(t,x)\psi(t,x)\right)\right] dx \, dt = 0.$$

Consider a $T > 0$ and a function $\Psi \in \mathcal{D}(\mathbb{R}^d)$. Thanks to the previous step, we can find a solution $\varphi \in L^\infty\left((-T,T) \times \mathbb{R}^d\right)$, $\varphi \in L^\infty\left((-T,T); L^1(\mathbb{R}^d)\right)$ to

$$\begin{cases} \frac{\partial}{\partial t}\varphi(t,x) + \mathrm{div}\left(b(t,x)\varphi(t,x)\right) = 0, t \in \mathbb{R}, \ x \in \mathbb{R}^d, \\[2mm] \varphi(t = T, x) = \Psi(x). \end{cases}$$

Then, using Lemma 6.1, we have

$$\frac{\partial}{\partial t}\varphi_\varepsilon(t,x) + \mathrm{div}\left(b(t,x)\varphi_\varepsilon(t,x)\right) = r_\varepsilon(t,x) \to 0 \quad in \ \ L^\infty\left((-T,T); L^1(\mathbb{R}^d)\right).$$

This function φ_ε is still not allowed as a test function because it does not have a compact support. The truncation in space is a simple matter and we skip this step. Time truncation can be achieved thanks to a nonincreasing function $K_\eta(t) = 1$ for $0 \le t \le T - \eta$, and $K_\eta(t) = 0$ for $t \ge T$. Then we use the test function $\psi = K_\eta(t)\varphi_\varepsilon$ in the weak formulation and arrive at

$$\int_0^\infty \int_{\mathbb{R}^d} u(t,x) \left[K_\eta(t)\frac{\partial \varphi_\varepsilon(t,x)}{\partial t} + \varphi_\varepsilon \frac{\partial K_\eta(t)}{\partial t} + K_\eta(t)\mathrm{div}\left(b(t,x)\varphi_\varepsilon(t,x)\right)\right] dx \, dt = 0,$$

$$\int_0^\infty \int_{\mathbb{R}^d} u(t,x) \left[K_\eta(t)r_\varepsilon(t,x) + \varphi_\varepsilon \frac{\partial K_\eta(t)}{\partial t}\right] dx \, dt = 0.$$

As $\varepsilon \to 0$, we deduce

$$\int_0^\infty \int_{\mathbb{R}^d} u(t,x)\varphi \frac{\partial}{\partial t} K_\eta(t) dx\, dt = 0.$$

But $\frac{\partial}{\partial t} K_\eta(t) \to -\delta(t = T)$ as $\eta \to 0$ and we deduce, that $\int_{\mathbb{R}^d} u(T,x)\Psi(x) = 0$, for a.e. $T > 0$. A similar argument for negative times concludes that $u \equiv 0$ and uniqueness is proved. $\qquad\square$

Exercise. Explain how Peano's non-uniqueness phenomena is handled in the above theory, for $b(x) = \sqrt{|x|}$ in one space dimension.

6.3 Generalized relative entropy: finite dimensional systems

We begin with describing the General Entropy Inequality in the case of matrices and we deal with two theories where it applies to give an entropy based understanding of time relaxation. In the framework of the Perron–Frobenius eigenvalue theorem it explains why the associated dynamic converges to the first (positive) eigenvector (once correctly normalized). In the framework of Floquet's eigenvalue theorem it explains why the associated dynamic converges to the (positive) periodic solution (once correctly normalized).

6.3.1 The Perron–Frobenius theorem

Let $a_{ij} > 0$, $1 \leq i,j \leq d$, be the coefficients of a matrix $A \in M_{d \times d}(\mathbb{R})$ (there are interesting issues with the case $a_{ij} \geq 0$ but we try to keep simplicity here). The Perron-Frobenius theorem (see [210] for instance) tells us that A has a first eigenvalue $\lambda > 0$ associated with a positive right eigenvector $N \in \mathbb{R}^d$, and a positive left eigenvector $\phi \in \mathbb{R}^d$,

$$\begin{cases} A.N = \lambda N, & N_i > 0 \quad \text{for} \quad i = 1,\ldots,d, \\[2mm] \phi.A = \lambda\phi, & \phi_i > 0 \quad \text{for} \quad i = 1,\ldots,d. \end{cases}$$

For later purposes, it is convenient to normalize these vectors, so that they are now uniquely defined. We choose

$$\sum_{i=1}^d N_i = 1, \qquad \sum_{i=1}^d N_i\, \phi_i = 1.$$

We set $\widetilde{A} = A - \lambda Id$ and consider the evolution equation

$$\frac{d}{dt} n(t) = \widetilde{A}.n(t), \qquad n(0) = n^0. \tag{6.12}$$

The solutions to this system converge as $t \to \infty$ with an exponential rate. Indeed, the following result is classical.

Proposition 6.5. *For positive matrices A and solutions to the differential system (6.12), we have,*

$$\rho := \sum_{i=1}^{d} \phi_i n_i(t) = \sum_{i=1}^{d} \phi_i n_i^0, \tag{6.13}$$

$$\sum_{i=1}^{d} \phi_i |n_i(t)| \leq \sum_{i=1}^{d} \phi_i |n_i^0|, \tag{6.14}$$

$$\underline{C}N_i \leq n_i(t) \leq \overline{C}N_i \quad \text{with constants given by} \quad \underline{C}N_i \leq n_i^0 \leq \overline{C}N_i, \tag{6.15}$$

and there is a constant $\alpha > 0$ such that, with ρ given in (6.13), we have

$$\sum_{i=1}^{d} \phi_i N_i \Big(\frac{n_i(t) - \rho N_i}{N_i} \Big)^2 \leq \sum_{i=1}^{d} \phi_i N_i \Big(\frac{n_i^0 - \rho N_i}{N_i} \Big)^2 e^{-\alpha t}. \tag{6.16}$$

Here, we wish to justify it with an entropy inequality.

Proposition 6.6. *Let $H(\cdot)$ be a convex function on $\mathbb{R}$, then the solution to (6.12) satisfies*

$$\frac{d}{dt} \sum_{i=1}^{d} \phi_i N_i H\Big(\frac{n_i(t)}{N_i} \Big)$$

$$= \sum_{i,j=1}^{d} \phi_i a_{ij} N_j \left[H'\Big(\frac{n_i(t)}{N_i} \Big) \Big[\frac{n_j(t)}{N_j} - \frac{n_i(t)}{N_i} \Big] - H\Big(\frac{n_j(t)}{N_j} \Big) + H\Big(\frac{n_i(t)}{N_i} \Big) \right]$$

$$\leq 0.$$

Definition 6.2. We call the quantity $\displaystyle\sum_{i=1}^{d} \phi_i N_i H\Big(\frac{n_i(t)}{N_i} \Big)$ General Relative Entropy.

Proof of Proposition 6.6. We denote by $\widetilde{a_{ij}}$ the coefficients of the matrix $\widetilde{A}$ and compute

$$\frac{d}{dt} \sum_i \phi_i N_i H\Big(\frac{n_i(t)}{N_i} \Big) = \sum_{i,j} \phi_i H'\Big(\frac{n_i(t)}{N_i} \Big) \widetilde{a_{ij}} n_j(t)$$
$$= \sum_{i,j} \phi_i \widetilde{a_{ij}} N_j H'\Big(\frac{n_i(t)}{N_i} \Big) \Big[\frac{n_j(t)}{N_j} - \frac{n_i(t)}{N_i} \Big],$$

because the additional $\frac{n_i(t)}{N_i}$ term vanishes since $\widetilde{A}.N = 0$. But we also have, again thanks to the equation on N and ϕ, that

$$\sum_{i,j} \phi_i \widetilde{a_{ij}} N_j \Big[H\Big(\frac{n_j(t)}{N_j} \Big) - H\Big(\frac{n_i(t)}{N_i} \Big) \Big] = 0.$$

Combining these two identities, we arrive at the equality in Proposition 6.6. The inequality follows because only the coefficients out of the diagonal, that satisfy $\widetilde{a_{ij}} = a_{ij} \geq 0$, enter here. $\qquad\square$

Proof of Proposition 6.5. Notice that, as a special case of H in Proposition 6.6, we can choose $H(u) = u$, which being convex together with $-H$ gives the equality

$$\frac{d}{dt} \sum_{i=1}^{d} \phi_i n_i(t) = 0.$$

And (6.13) follows. In particular this identifies the value ρ mentioned in (6.13).

The second statement (6.14) follows immediately by choosing the (convex) entropy function $H(u) = |u|$.

As for the third statement (6.15), let us consider for instance the upper bound. It follows choosing the (convex) entropy function $H(u) = (u - \overline{C})_+^2$ because for this nonnegative function we have

$$\sum_{i=1}^{d} \phi_i N_i H\left(\frac{n_i^0}{N_i}\right) = 0.$$

Therefore, because the General Relative Entropy decays, it remains zero for all times,

$$\sum_{i=1}^{d} \phi_i N_i H\left(\frac{n_i(t)}{N_i}\right) = 0,$$

which proves the result.

It remains to prove the exponential time decay statement (6.16). To do that, we work on

$$h(t, x) = n(t, x) - \rho N,$$

which verifies $\int \varphi[n(t, x) - \rho N] dx = 0$ and satisfies the same equation as n. Then, we use the quadratic entropy function $H(u) = u^2$ and the General Entropy Inequality gives

$$\frac{d}{dt} \sum_{i=1}^{d} \phi_i N_i \left(\frac{h_i(t)}{N_i}\right)^2 = -\sum_{i,j=1}^{d} \phi_i a_{ij} N_j \left(\frac{h_j(t)}{N_j} - \frac{h_i(t)}{N_i}\right)^2 \leq 0.$$

Then, we need a discrete Poincaré inequality.

Lemma 6.2. *Being given* $\phi_i > 0$, $N_i > 0$, $a_{ij} > 0$ *for* $i = 1, \ldots, d$, $j = 1, \ldots, d$, $i \neq j$, *there is a constant* $\alpha > 0$ *such that for all vector* m *of components* m_i, $1 \leq i \leq d$ *satisfying*

$$\sum_{i=1}^{d} \phi_i m_i = 0,$$

we have (Poincaré inequality)

$$\sum_{i,j=1}^{d} \phi_i a_{ij} N_j \left(\frac{m_j}{N_j} - \frac{m_i}{N_i}\right)^2 \geq \alpha \sum_{i=1}^{d} \phi_i N_i \left(\frac{m_i}{N_i}\right)^2.$$

With this lemma, we conclude

$$\frac{d}{dt} \sum_{i=1}^{d} \phi_i N_i \left(\frac{h_i(t)}{N_i}\right)^2 \leq -\alpha \sum_{i=1}^{d} N_i \left(\frac{h_i(t)}{N_i}\right)^2,$$

and then, (6.16) follows by a simple use of Gronwall lemma.   □

Proof of Lemma 6.2. After renormalizing the vector m (when it does not vanish, otherwise the result is obvious), we may suppose that

$$\sum_{i=1}^{d} \phi_i m_i = 0, \qquad \sum_{i=1}^{d} \phi_i N_i \left(\frac{m_i}{N_i}\right)^2 = 1.$$

Then we argue by contradiction. If such an α does not exist, this means that we can find a sequence of vectors $(m^k)_{(k \geq 1)}$ such that

$$\sum_{i=1}^{d} \phi_i m_i^k = 0, \qquad \sum_{i=1}^{d} \phi_i N_i \left(\frac{m_i^k}{N_i}\right)^2 = 1, \qquad \sum_{i,j=1}^{d} \phi_i a_{ij} N_j \left(\frac{m_j^k}{N_j} - \frac{m_i^k}{N_i}\right)^2 \leq 1/k.$$

After extraction of a subsequence, we may pass to the limit $m^k \to \bar{m}$ and this vector satisfies

$$\sum_{i=1}^{d} \phi_i \bar{m}_i = 0, \qquad \sum_{i=1}^{d} \phi_i N_i \left(\frac{\bar{m}_i}{N_i}\right)^2 = 1,$$

$$\sum_{i,j=1}^{d} \phi_i a_{ij} N_j \left(\frac{\bar{m}_j}{N_j} - \frac{\bar{m}_i}{N_i}\right)^2 = 0.$$

Therefore, from this last relation, for all i and $j = 1, \ldots, d$, we have

$$\frac{\bar{m}_i}{N_i} = \frac{\bar{m}_j}{N_j} := \nu.$$

By the zero sum condition, we have $\nu = 0$ because

$$\nu \sum_{i=1}^{d} \phi_i = 0.$$

In other words, $\bar{m} = 0$ which contradicts the normalization and thus such a α should exist.   □

Remark 6.2. 1. This entropy structure is related to a characterization of M-matrices, i.e., those whose terms out of the diagonal are negative, diagonal terms are positive and dominate the corresponding line. Such a matrix has an inverse with positive coefficients. It was noticed in [216] that a characterization of M-matrices uses the existence of positive eigenvectors as N and ϕ above. Let us point out that the General Relative Entropy inequality also holds for M-matrices because the diagonal terms do not appear in the inequality of Lemma 6.2.

2. The matrix with (positive) coefficients $b_{ij} = \phi_i \, a_{ij} \, N_j$ is doubly stochastic, i.e., the sum of the lines and columns is 1 (see for instance [210]).

3. Notice that $a_{ii} - \lambda < 0$ because $\sum_j \widetilde{a_{ij}} N_j = 0$. Therefore the matrix C with coefficients $c_{ij} = \frac{1}{N_i} \widetilde{a_{ij}} \, N_j$ is that of a Markov process. In other words, we set $y_i = x_i / N_i$, then it satisfies

$$\frac{d}{dt} y_i(t) = c_{ij} y_j(t),$$

and the vector $(1, 1, \ldots, 1)$ is the (positive) eigenvector associated to the eigenvalue 0 of the matrix C, i.e., $c_{ii} = \sum_{j \neq i} c_{ij}$ and $c_{ij} \geq 0$. Then, $(N_i \phi_i)_{(i=1,\ldots,d)}$ is the invariant measure of the Markov process. In particular this explains the entropy property which is classical for Markov processes ([224]).

6.3.2 The Floquet theory

We now consider T-periodic coefficients $a_{ij}(t) > 0$, $1 \leq i, j \leq d$, i.e., $a_{ij}(t + T) = a_{ij}(t)$. And we denote by $A(t) \in M_{d \times d}$ the corresponding matrix. Again our motivation comes from several questions in biology where such structures arise, such as seasonal rhythm, circadian rhythm, see [120, 63, 109, 19] for instance.

The Floquet theorem tells us that there is a first 'Floquet eigenvalue' $\lambda_{per} > 0$ and two positive T-periodic functions $N(t) \in \mathbb{R}^d$, $\phi(t) \in \mathbb{R}^d$ that are periodic solutions (uniquely defined up to multiplication by a constant) to the differential systems

$$\frac{d}{dt} N(t) = [A(t) - \lambda_{per} Id].N(t), \qquad (6.17)$$

$$\frac{d}{dt} \phi(t) = \phi(t).[A(t) - \lambda_{per} Id]. \qquad (6.18)$$

Up to a normalization, these elements $(\lambda_{per} > 0, N(t) > 0, \phi(t))$ are unique and we normalize again as

$$\int_0^T \sum_{i=1}^d N_i(t) dt = 1, \qquad \int_0^T \sum_{i=1}^d \phi_i(t) N_i(t) dt = 1.$$

We recall that this case of Floquet theory (which applies to more general situations than positive matrices) is an application of the Perron–Frobenius theorem to the resolvent matrix

$$S(t) = e^{\int_0^t A(s)ds},$$

which has positive coefficients also. A classical introduction to the subject can be found in [65].

Again, we set $\widetilde{A(t)} = A(t) - \lambda_{per} Id$ and consider the differential system

$$\frac{d}{dt}n(t) = \tilde{A}.n(t), \qquad n(0) = n^0.$$

In the present context we obtain the following version of Proposition 6.5.

Proposition 6.7. *For positive matrices A we have,*

$$\rho := \sum_{i=1}^{d} \phi_i(t)n_i(t) = \sum_{i=1}^{d} \phi_i(t=0)n_i^0, \tag{6.19}$$

$$\sum_{i=1}^{d} \phi_i(t)|n_i(t)| \le \sum_{i=1}^{d} \phi_i(t=0)|n_i^0|; \tag{6.20}$$

if for some constants, we have $\underline{C}N_i(t=0) \le n_i^0 \le \overline{C}N_i(t=0)$, then

$$\underline{C}N_i(t) \le n_i(t) \le \overline{C}N_i(t), \tag{6.21}$$

and there is a constant $\alpha > 0$ such that

$$\sum_{i=1}^{d} \phi_i(t)N_i(t)\left(\frac{n_i(t)-\rho N_i(t)}{N_i(t)}\right)^2 \le \sum_{i=1}^{d} \phi_i^0 N_i^0 \left(\frac{n_i^0 - \rho N_i^0}{N_i^0}\right)^2 e^{-\alpha t}. \tag{6.22}$$

Again, this can be justified thanks to entropy inequalities.

Proposition 6.8. *Let $H(\cdot)$ be a convex function on $\mathbb{R}$, then we have*

$$\frac{d}{dt}\sum_{i=1}^{d}\phi_i(t)N_i(t)H\left(\frac{n_i(t)}{N_i(t)}\right)$$

$$= \sum_{i,j=1}^{d}\phi_i a_{ij}N_j\left[H'\left(\frac{n_i}{N_i}\right)\left[\frac{n_j}{N_j}-\frac{n_i}{N_i}\right]-H\left(\frac{n_j}{N_j}\right)+H\left(\frac{n_i}{N_i}\right)\right]$$

$$\le 0.$$

These two propositions are variants of the corresponding ones in the Perron–Frobenius theorem and we leave the proofs to the reader. Adapting Lemma 6.2 requires an additional compactness argument based on the Ascoli–Arzela Theorem.

6.4 Generalized relative entropy: parabolic and integral PDEs

We now explain the notion of General Relative Entropy on continuous models. We begin with the most classical equation, namely the parabolic equation for the unknown $n(t, x)$,

$$\frac{\partial n}{\partial t} - \sum_{i,j=1}^{d} \frac{\partial}{\partial x_i}\left(a_{ij}\frac{\partial n}{\partial x_j}\right) + \sum_{i=1}^{d} \frac{\partial}{\partial x_i}(b_i n) + dn = 0, \qquad x \in \mathbb{R}^d, \tag{6.23}$$

where the coefficients depend on t and x, $d \equiv d(t, x)$ (no sign assumed), $b_i \equiv b_i(t, x)$, and the symmetric matrix $A(t, x) = \left(a_{ij}(t, x)\right)_{1 \leq i,j \leq d}$ satisfies $A(t, x) \geq 0$. We could possibly set the equation on a domain and assume Dirichlet, zero-flux, mixed or periodic boundary conditions and then include them in the above calculation.

Here, it is not obvious to derive a priori bounds on the solution $n(t, x)$, in contrast to the case $A \geq \nu Id > 0$, $b_i \equiv 0$, $d(x) \geq 0$ where we have, multiplying the equation by $n|n|^{p-2}$ with $p > 1$,

$$\frac{d}{dt} \int \frac{|n(t, x)|^p}{p} dx + \frac{4\nu(p-1)}{p^2} \int |\nabla n^{p/2}|^2 dx \leq 0.$$

Indeed the only remarkable property of (6.23) is the mass conservation and L^1 contraction principle

$$\frac{d}{dt} \int n(t, x) dx + \int d(t, x) n(t, x) dx = 0,$$

$$\frac{d}{dt} \int \left(n(t, x)\right)_+ dx + \int d(t, x)\left(n(t, x)\right)_+ dx \leq 0.$$

On the other hand the conservative Fokker-Planck equation is very standard when $b = -\nabla V$ for some convex potential with enough growth at infinity

$$\frac{\partial n}{\partial t} - \Delta n - \mathrm{div}(\nabla V\, n) = 0.$$

Then, one has $N = e^{-V}$ and the relative entropy $\int n \ln\left(\frac{n}{N}\right) dx$ is a standard object related to log-Sobolev inequalities, [12, 52, 155, 228]. It decays with time. Of course, here we still have the family $\int NH\left(\frac{n}{N}\right) dx$ of relative entropies. All of these entropies, for all convex functions $H(\cdot)$, decays in time, and not only $H(u) = u \ln(u)$.

6.4.1 Coefficients independent of time

In the case of coefficients independent of time, and depending on the values of $a_{ij}(x)$, $b(x)$ and $d(x)$, the solution can exhibit exponential growth or decay as

$t \to \infty$. Therefore, we will assume that 0 is the first eigenvalue and, following the Krein–Rutman theorem (see [72]), we also assume that we can find two functions $N(x) > 0$, $\phi(x) > 0$, such that

$$
\begin{cases}
-\displaystyle\sum_{i,j=1}^{d} \frac{\partial}{\partial x_i}\left(a_{ij}(x)\frac{\partial N}{\partial x_j}\right) + \sum_{i=1}^{d} \frac{\partial}{\partial x_i}\left(b_i(x)N\right) + d(x)N = 0, \\[4mm]
N(x) > 0, \qquad \int N(x)dx = 1,
\end{cases}
\tag{6.24}
$$

$$
\begin{cases}
-\displaystyle\sum_{i,j=1}^{d} \frac{\partial}{\partial x_i}\left(a_{ij}(x)\frac{\partial \phi}{\partial x_j}\right) - \sum_{i=1}^{d} b_i(x)\frac{\partial}{\partial x_i}\phi + d(x)\phi = 0, \\[4mm]
\phi(x) > 0, \qquad \int N(x)\phi(x)dx = 1.
\end{cases}
\tag{6.25}
$$

These are the first eigenvectors; N for the direct problem and ϕ for the dual operator. Notice that such eigenelements do not always exist but there are standard examples, namely when $d \equiv 0$, $A = Id$ and there is a potential V such that $b = -\nabla V$. Then, one can readily check that solutions to (6.24)–(6.25) are

$$
N = e^{-V} \qquad \phi \equiv 1,
$$

when $V(x) \to \infty$ as $|x| \to \infty$ fast enough in order to fulfill the integrability conditions.

The general relative entropy property of the parabolic equation (6.23) can be expressed as

Lemma 6.3. *For coefficients independent of t, assume that there exist eigenelements N, ϕ satisfying (6.24)–(6.25). Then for all convex functions $H : \mathbb{R} \to \mathbb{R}$, and all solutions n to (6.23) with sufficient decay in x to zero at infinity ($|n^0| \le CN$), we have*

$$
\frac{d}{dt}\int \phi(x)\, N(x)\, H\!\left(\frac{n(t,x)}{N(x)}\right)dx
$$

$$
= -\int \phi\, N\, H''\!\left(\frac{n(t,x)}{N(x)}\right) \sum_{i,j=1}^{d} a_{ij}\frac{\partial}{\partial x_i}\!\left(\frac{n}{N}\right)\frac{\partial}{\partial x_j}\!\left(\frac{n}{N}\right)\, dx \le 0.
$$

For conservative equations, i.e., $d \equiv 0$, it is usual to take $\phi \equiv 1$, and then the corresponding principle is classical (especially related to stochastic differential equations and Markov processes, [224]).

Proof. We just calculate (leaving the intermediary steps to the reader)

$$
\frac{\partial}{\partial t}\!\left(\frac{n}{N}\right) - \sum_{i,j=1}^{d} \frac{\partial}{\partial x_i}\!\left[a_{ij}\frac{\partial}{\partial x_j}\!\left(\frac{n}{N}\right)\right] + 2N\sum_{i,j=1}^{d} a_{ij}\frac{\partial}{\partial x_i}\!\left(\frac{n}{N}\right)\frac{\partial}{\partial x_j}\!\left(\frac{1}{N}\right) + b\cdot\nabla\!\left(\frac{n}{N}\right) = 0.
$$

Therefore, for any smooth function H, we arrive at

$$\frac{\partial}{\partial t}H\Big(\frac{n}{N}\Big) - \sum_{i,j=1}^{d}\frac{\partial}{\partial x_i}\Big[a_{ij}\frac{\partial}{\partial x_j}H\Big(\frac{n}{N}\Big)\Big] + 2N\sum_{i,j=1}^{d}a_{ij}\frac{\partial}{\partial x_i}H\Big(\frac{n}{N}\Big)\frac{\partial}{\partial x_j}\Big(\frac{1}{N}\Big)$$

$$+ b\cdot\nabla H\Big(\frac{n}{N}\Big) + H''\Big(\frac{n}{N}\Big)\sum_{i,j=1}^{d}a_{ij}\frac{\partial}{\partial x_i}\Big(\frac{n}{N}\Big)\frac{\partial}{\partial x_j}\Big(\frac{n}{N}\Big) = 0.$$

At this stage we can 'undo' the calculation that led from an equation on n to an equation on n/N and we arrive at

$$\frac{\partial}{\partial t}NH\Big(\frac{n}{N}\Big) - \sum_{i,j=1}^{d}\frac{\partial}{\partial x_i}\Big[a_{ij}\frac{\partial}{\partial x_j}NH\Big(\frac{n}{N}\Big)\Big] + NH''\Big(\frac{n}{N}\Big)\sum_{i,j=1}^{d}a_{ij}\frac{\partial}{\partial x_i}\Big(\frac{n}{N}\Big)\frac{\partial}{\partial x_j}\Big(\frac{n}{N}\Big)$$

$$+ \sum_{i=1}^{d}\frac{\partial}{\partial x_i}\Big[b_i NH\Big(\frac{n}{N}\Big)\Big] + dNH\Big(\frac{n}{N}\Big) = 0.$$

Finally, combining it with the equation on ϕ, we deduce that

$$\frac{\partial}{\partial t}\phi NH\Big(\frac{n}{N}\Big) - \sum_{i,j=1}^{d}\frac{\partial}{\partial x_i}\Big[\phi a_{ij}\frac{\partial}{\partial x_j}NH\Big(\frac{n}{N}\Big)\Big] + \sum_{i,j=1}^{d}\frac{\partial}{\partial x_i}\Big[a_{ij}NH\Big(\frac{n}{N}\Big)\frac{\partial}{\partial x_j}\phi\Big]$$

$$+ \phi NH''\Big(\frac{n}{N}\Big)\sum_{i,j=1}^{d}a_{ij}\frac{\partial}{\partial x_i}\Big(\frac{n}{N}\Big)\frac{\partial}{\partial x_j}\Big(\frac{n}{N}\Big) = 0.$$

After integration in x (because we have assumed sufficient decay in x to zero at infinity), we arrive at the result stated in Lemma 6.3. $\qquad\square$

This lemma can be used in the directions indicated in Section 6.3 (a priori estimates, long time convergence to a steady state) and we refer to [177, 175, 176] for specific examples.

As far as long time convergence is concerned, we notice that, as in Lemma 6.2, a control of entropy by entropy dissipation is useful for exponential convergence as $t \to \infty$ as in (6.16). For the quadratic entropy, this follows from the Poincaré inequality

$$\nu\int\phi N\Big(\frac{m}{N}\Big)^2 \leq 2\int\phi N\sum_{i,j=1}^{d}a_{ij}\frac{\partial}{\partial x_i}\Big(\frac{m}{N}\Big)\frac{\partial}{\partial x_j}\Big(\frac{m}{N}\Big), \quad \text{when} \int\phi m = 0.$$

Such inequalities, as well as log-Sobolev inequalities, are classical when $N = e^{-V}$ for a potential $V(x)$ with superlinear growth at infinity ([155] for this result and [228] for general issues on this subject). The change of unknown function to $n\phi$ and $N\phi$ gives the condition $N\phi = e^{-V}$ for $V(x)$ with superlinear growth to ensure the Poincaré inequality. We are not aware of any general condition on d, b and A in this direction.

6.4.2   Time dependent coefficients

In fact the above manipulations are also valid for time dependent coefficients. A situation similar to the Floquet theory and which is therefore useful for periodic coefficients for instance. We now consider solutions to

$$
\left\{
\begin{aligned}
&\frac{\partial}{\partial t}N(t,x) - \sum_{i,j=1}^{d} \frac{\partial}{\partial x_i}\Big(a_{ij}(x)\frac{\partial N}{\partial x_j}\Big) + \sum_{i=1}^{d} \frac{\partial}{\partial x_i}\big(b_i(x)N\big) + d(x)N = 0,\\[2mm]
&\hspace{7cm} N(t,x) > 0,
\end{aligned}
\right.
\tag{6.26}
$$

$$
\left\{
\begin{aligned}
&\frac{\partial}{\partial t}\phi(t,x) - \sum_{i,j=1}^{d} \frac{\partial}{\partial x_i}\Big(a_{ij}(x)\frac{\partial \phi}{\partial x_j}\Big) - \sum_{i=1}^{d} b_i(x)\frac{\partial}{\partial x_i}\phi + d(x)\phi = 0,\\[2mm]
&\hspace{7cm} \phi(t,x) > 0.
\end{aligned}
\right.
\tag{6.27}
$$

Then we have

Lemma 6.4. *For all convex functions $H : \mathbb{R} \to \mathbb{R}$, and all solutions n to (6.23) with sufficient decay in x to zero at infinity, we have*

$$
\frac{d}{dt}\int \phi(t,x)\, N(t,x)\, H\Big(\frac{n(t,x)}{N(t,x)}\Big)dx
$$

$$
= -\int \phi\, N\, H''\Big(\frac{n(t,x)}{N(t,x)}\Big) \sum_{i,j=1}^{d} a_{ij}\,\frac{\partial}{\partial x_i}\Big(\frac{n}{N}\Big)\,\frac{\partial}{\partial x_j}\Big(\frac{n}{N}\Big)\, dx \le 0.
$$

Again we leave the proof of this variant to the reader.

6.4.3   Scattering equations

To exhibit another class of equation where the General Relative Entropy inequality holds true, let us mention the scattering (also called linear Boltzmann) equation

$$
\frac{\partial}{\partial t}n(t,x) + k_T(x)\, n(t,x) = \int_{\mathbb{R}^d} k(x,y)\, n(t,y)\, dy.
\tag{6.28}
$$

Here we restrict ourselves to coefficients independent of time for simplicity, but the same inequality holds in the time dependent case as before. We assume that

$$
0 \le k_T(\cdot) \in L^\infty(\mathbb{R}^d), \qquad 0 \le k(x,y) \in L^1 \cap L^\infty(\mathbb{R}^{2d}).
$$

And we do not make a special assumption on the symmetry of the cross-section $k(x,y)$ as motivated by turning kernels that appear in various applications such as bacterial movement [5, 75, 54, 189] and Section 5.6.

Again, changing k_T in $k_T + \lambda$ if necessary in order to have a zero first eigenvalue, we assume that there are solutions $N(x)$ and $\phi(x)$ to the stationary equation and its adjoint, namely

$$k_T(x)\, N(x) = \int_{\mathbb{R}^d} k(x,y)\, N(y)\, dy, \qquad N(x) > 0, \qquad (6.29)$$

$$k_T(x)\, \phi(x) = \int_{\mathbb{R}^d} k(y,x)\, \phi(y)\, dy, \qquad \phi(x) > 0. \qquad (6.30)$$

These two steady state solutions allow us to derive the General Relative Entropy inequality

Lemma 6.5. *With the above notation, we have*

$$\frac{\partial}{\partial t}\left[\phi(x)\, N(x)\, H\!\left(\frac{n(x)}{N(x)}\right)\right]$$

$$+ \int_{\mathbb{R}^d} k(x,y)\left[\phi(y)N(x)H\!\left(\frac{n(t,x)}{N(x)}\right) - \phi(x)N(y)H\!\left(\frac{n(t,y)}{N(y)}\right)\right]dy$$

$$= \int k(x,y)\phi(x)N(y)\left[H\!\left(\frac{n(t,x)}{N(x)}\right) - H\!\left(\frac{n(t,y)}{N(y)}\right)\right.$$

$$\left. + H'\!\left(\frac{n(t,x)}{N(x)}\right)\left[\frac{n(t,y)}{N(y)} - \frac{n(t,x)}{N(x)}\right]\right]dy,$$

and also (after integration in x),

$$\frac{d}{dt}\int_{\mathbb{R}^d}\left[\phi(x)\, N(x)\, H\!\left(\frac{n(x)}{N(x)}\right)\right]$$

$$= \int k(x,y)\phi(x)N(y)\left[H\!\left(\frac{n(t,x)}{N(x)}\right) - H\!\left(\frac{n(t,y)}{N(y)}\right)\right.$$

$$\left. + H'\!\left(\frac{n(t,x)}{N(x)}\right)\left[\frac{n(t,y)}{N(y)} - \frac{n(t,x)}{N(x)}\right]\right]dy$$

$$\leq 0.$$

Again we leave to the reader the easy computation that leads to this result and just indicate a class of classical examples where N and ϕ can be computed explicitly.

Example 1. We consider the case where the cross-section in the scattering equation is given by

$$k(x,y) = k_1(x)k_2(y).$$

Then we arrive at (up to a multiplicative constant)

$$N(x) = \frac{k_1(x)}{k_T(x)}, \qquad \phi(x) = \frac{k_2(x)}{k_T(x)},$$

and we need the compatibility condition

$$\int_{\mathbb{R}^d} \frac{k_2(x)k_1(x)}{k_T(x)^2} dx = 1.$$

As in the case of the Perron–Frobenius thorem in Section 6.3.1, this means that 0 is the first eigenvalue, a condition that can always be met, changing if necessary k_T in $\lambda + k_T$.

Example 2. We consider the more general case where there exists a function $N(x) > 0$ such that the scattering cross-section satisfies the symmetry condition (usually called detailed balance or microreversibility)

$$k(y,x)N(x) = k(x,y)N(y).$$

Then the choice $k_T(y) = \int_{\mathbb{R}^d} k(x,y)dx$ gives the solutions $N(x)$ to (6.29), and $\phi(x) = 1$ to equation (6.30).

Again we conclude with long time convergence and the possibility to prove exponential time decay to the steady state. As in Lemma 6.2, this follows from a control of entropy by entropy dissipation and thus for the quadratic entropy, from the Poincaré inequality

$$\nu \int \phi(x)N(x)\Big(\frac{h}{N}\Big)^2 dx \leq \int_{\mathbb{R}^d} k(y,x)\phi(x)N(y)\left[\frac{h(x)}{N(x)} - \frac{h(y)}{N(y)}\right]^2 dy\, dx,$$

whenever

$$\int_{\mathbb{R}^d} \phi(x)h(x)dx = 0.$$

This is not always true but holds whenever there is a function $\psi > 0$ such that

$$\nu_1 = \int N\phi^2/\psi < \infty, \qquad \nu_2\psi(y)N(x) \leq k(x,y), \qquad \nu = (\nu_1\nu_2)^{-1},$$

a condition that is fulfilled for instance if we work on a bounded domain in velocity and k positive (the difficulties in practical examples such as the cell division equation is that ϕ need not be bounded in unbounded domains and N can vanish at infinity).

We write, for any function $\psi > 0$, and $\nu_1 = \int N/\psi$,

$$\int \phi(x)N(x)\Big(\frac{h}{N}(x)\Big)^2 dx = \int \phi(x)N(x) \Big(\int [\frac{h}{N}(x) - \frac{h}{N}(y)]\phi(y)N(y)dy\Big)^2 dx$$
$$\leq \nu_1 \int \int \phi(x)N(x)[\frac{h}{N}(x) - \frac{h}{N}(y)]^2\psi(y)N(y)dydx$$
$$\leq \nu_1\nu_2 \int \int \phi(x)k(x,y)N(y)[\frac{h}{N}(x) - \frac{h}{N}(y)]^2dydx.$$

Notice that a large class of the examples above enter this condition but not with the choice $\psi = \phi$.

6.5   BV functions, Sobolev imbeddings

In this section we gather several results that have been used in several chapters and that are based on the theory of functional spaces and related inequalities.

6.5.1   BV space; limits

In Sections 1.6.1 and 2.2, we have used an elementary property of functions with bounded variations (for which a complete and elaborate mathematical theory exists, see [8]).

Lemma 6.6. *Let* $n \in C^1(\mathbb{R}^+)$ *satisfy* $\int_0^\infty |\frac{dn(t)}{dt}|\, dt := \|n\|_{TV} < \infty$, *then* $n(t)$ *admits a limit* L *as* $t \to \infty$.

The notation $\|n\|_{TV}$ holds for Total Variation, a seminorm that allows us to define the BV space of L^1 functions with finite TV seminorms.

Proof. We first notice that the assumption implies that

$$\int_A^\infty |\frac{dn(t)}{dt}|\, dt = \|n\|_{TV} - \int_0^A |\frac{dn(t)}{dt}|\, dt \to 0, \qquad \text{as } A \to \infty.$$

Therefore, we have, for $p > k$ (integers),

$$|n(k) - n(p)| \le \int_k^p |\frac{dn(t)}{dt}|\, dt \to 0, \qquad \text{as } k \to \infty,$$

which implies that $\big(n(p)\big)_{p \ge 1}$ is a Cauchy sequence and thus admits a limit L.

It remains to write that, as $x \to \infty$,

$$|n(x) - L| = \lim_{p \to \infty} |n(x) - n(p)| \le \lim_{p \to \infty} \int_x^p |\frac{dn(t)}{dt}|\, dt = \int_x^\infty |\frac{dn(t)}{dt}|\, dt \to 0,$$

and the result is proved.   □

An extension is as follows (we write it in a general form but supposing the C^1 regularity is enough for the purpose mentioned above)

Lemma 6.7. *Let* $n \in L^\infty(\mathbb{R}^+)$ *satisfy* $\int_0^\infty (\frac{dn(t)}{dt})_+\, dt < \infty$, *then* $n(t)$ *is Total Variation bounded and thus admits a limit* L *as* $t \to \infty$.

Proof. We can write (still in the C^1 case to avoid the difficulty to justify it for distributional derivatives, which can be done for instance by density)

$$n(B) - n(A) = \int_A^B \frac{dn(t)}{dt}\, dt = \int_0^\infty (\frac{dn(t)}{dt})_+\, dt - \int_0^\infty (\frac{dn(t)}{dt})_-\, dt.$$

Therefore

$$\int_0^\infty (\frac{dn(t)}{dt})_-\, dt \le 2\|n\|_{L^\infty(\mathbb{R}^+)} + \int_0^\infty (\frac{dn(t)}{dt})_+\, dt.$$

This proves that $\int_0^\infty |\frac{dn(t)}{dt}|\, dt < \infty$, and thus we are reduced to the previous lemma.   □

6.5.2   Sobolev inequalities

(i) Sobolev inequality. The standard Sobolev inequality in $\mathbb{R}^d$ (see [98, 118] for instance) states that for all functions $u \in \mathcal{S}(\mathbb{R}^d)$ (smooth functions with fast decay at infinity)

$$\|u\|_{L^{p^*}(\mathbb{R}^d)} \le C(d,p)\|\nabla u\|_{L^p(\mathbb{R}^d)}, \quad \forall 1 \le p < d, \quad \frac{1}{p^*} = \frac{1}{p} - \frac{1}{d}. \tag{6.31}$$

The particular relation between p and q stands for homogeneity reasons in x. For $p = d$ the inequality (6.31) does not hold (with $r = \infty$) and variants in the space BMO (Bounded Mean Oscillations) can be stated, [157].

(ii) Gagliardo–Nirenberg–Sobolev inequality can be seen as an interpolation of this inequality as far as $p < d$ below: for some $C(d,p,q)$ we have

$$\begin{cases} \|u\|_{L^r(\mathbb{R}^d)} \le C\|\nabla u\|^{\theta}_{L^p(\mathbb{R}^d)}\|u\|^{1-\theta}_{L^q(\mathbb{R}^d)}, \\[2mm] \forall 1 \le p < d, \quad \forall 1 \le q \le \infty, \quad \frac{1}{r} = \frac{\theta}{p^*} + \frac{1-\theta}{q}. \end{cases} \tag{6.32}$$

The interesting fact here is that we can also reach the limiting case $p = d$: for some $C(d,q)$ we also have

$$\begin{cases} \|u\|_{L^r(\mathbb{R}^d)} \le C\|\nabla u\|^{\theta}_{L^d(\mathbb{R}^d)}\|u\|^{1-\theta}_{L^q(\mathbb{R}^d)}, \\[2mm] \forall 1 \le q < \infty, \quad \frac{1}{r} = \frac{1-\theta}{q}. \end{cases} \tag{6.33}$$

A special case is used in Section 5.2.2. It is derived with $u = \varrho^{\bar{p}/2}$, $\theta = \frac{\bar{p}}{\bar{p}+1}$, $p = 2$, $r = 2\frac{\bar{p}+1}{\bar{p}}$, $q = \frac{d}{\bar{p}}$. Then we arrive at

$$\int_{\mathbb{R}^d} \varrho^{p+1} dx \le C\|\varrho\|_{L^{d/2}(\mathbb{R}^d)} \int_{\mathbb{R}^d} |\nabla \varrho^{p/2}|^2 dx, \quad p+1 \ge \frac{d}{2}, \tag{6.34}$$

(after changing $\bar{p}$ in p).

6.5.3   Logarithmic inequalities

(i) Logarithmic Sobolev inequalities. There are several related results. As we saw, quantities like $u \ln(u)$ play a fundamental role in chemotaxis (see Section 5.2.1 for instance). Related are the logarithmic Sobolev inequalities (see [76] and the references therein). For

$$\int_{\mathbb{R}^d} |u|^p dx = 1, \quad 1 \le p < d,$$

we have

$$\int_{\mathbb{R}^d} |u|^p \ln|u| \, dx \le \frac{d}{p^2} \ln\left[\mathcal{L}_p \int_{\mathbb{R}^d} |\nabla u|^p dx\right],$$

and the best constant is known and attained by explicit functions.

(ii) Poincaré and Logarithmic Sobolev inequalities with weights. There are inequalities closer to what we encountered in Section 5.2.1. Namely, for $N(x) = e^{-(V_1+V_2)}$ with V_2 bounded and $D^2V_1 \geq C > 0$, as soon as the function u satisfies

$$\int_{\mathbb{R}^d} u(x)\, N(x)dx = 0,$$

we have the Poincaré inequality

$$\int_{\mathbb{R}^d} |u|^2\, N(x)\, dx \leq C \int_{\mathbb{R}^d} N(x)\, |\nabla u|^2 dx.$$

If u satisfies

$$\int_{\mathbb{R}^d} u(x)\, N(x)dx = 1,$$

we have the Logarithmic Sobolev inequality

$$\int_{\mathbb{R}^d} |u| \ln |u(x)|\, N(x)\, dx \leq C \int_{\mathbb{R}^d} N(x)\, |\nabla \sqrt{u}|^2 dx.$$

For proofs and a more precise account on this subject, we refer to [155, 228] for instance.

(iii) The logarithmic Hardy–Littlewood–Sobolev inequality (see [50, 21, 157]) is well adapted to the free energy of the Keller–Segel system.

Lemma 6.8. [50, 21] *Let f be a nonnegative function in $L^1(\mathbb{R}^d)$ such that $f \log f$ belongs to $L^1(\mathbb{R}^d)$. Set $\int_{\mathbb{R}^d} f\, dx = M$, then*

$$\int_{\mathbb{R}^d} f \log f\, dx + \frac{d}{M} \int \int_{\mathbb{R}^d \times \mathbb{R}^d} f(x)f(y) \log|x-y|\, dx\, dy \geq M[\ln M - C(d)], \quad (6.35)$$

with $C(2) := 1 + \log \pi$.

This lemma can be understood in various ways. One possible connection is through the elliptic equation in dimension $d = 2$,

$$-\Delta c = f.$$

For $f \in L^1(\mathbb{R}^2)$, we do not have $c \in L^\infty(\mathbb{R}^2)$ (this is a limiting case), nor do we have $\nabla c \in L^2(\mathbb{R}^2)$. But for $f \ln f \in L^1(\mathbb{R}^d)$ this lemma asserts that $\int_{\mathbb{R}^2} |\nabla c|^2 dx = \int_{\mathbb{R}^2} f(x)\, c(x)\, dx$ is bounded.

It can also be considered as the dual of Moser–Trudinger type inequalities, see [50, 21, 157],

$$\int_\Omega e^{|h|} dx \leq C_\Omega \exp\left[\frac{1}{8\theta} \int_\Omega |\nabla h|^2 dx + |\Omega|^{-1} \left| \int_\Omega h dx \right| \right],$$

with θ the minimum interior angle of the piecewise smooth domain Ω and $h \in H^1(\Omega)$.

(iv) Logarithmic loss in limiting cases of the Young inequality. In Section 5.6.1, we used another logarithmic loss, this time in the limiting cases of the Young inequality. Consider a solution to

$$\Delta u(x) = n(x), \qquad x \in \mathbb{R}^d.$$

The solution is given by a singular convolution and we are interested in the limiting case when $p = 1$ of the Young inequality

$$\|\nabla u\|_{L^{p^*}(\mathbb{R}^d)} \leq C(d,p) \, \|n\|_{L^p(\mathbb{R}^d)} \quad \forall 1 < p < d, \qquad \frac{1}{p^*} = \frac{1}{p} - \frac{1}{d}.$$

To go further and treat $p = 1$, we decompose the gradient in long and short range parts

$$\nabla u = \nabla u_L + \nabla u_S.$$

Then the long range interaction part enjoys good properties for our purpose,

$$\nabla u_L = \lambda_d \, \frac{x \, \mathbf{1}_{\{|x|>1\}}}{|x|^d} * n, \qquad \|\nabla u_L\|_\infty \leq C\|n\|_1.$$

As for the short range part we write

$$\nabla u_S = \lambda_d \, \frac{x \, \mathbf{1}_{\{|x|<1\}}}{|x|^d} * n.$$

Lemma 6.9. *For $p > 1$ one has*

$$\|\nabla u_S\|_{L^{d/(d-1)}(\mathbb{R}^d)} \leq C(d,p)\|n\|_{L^1(\mathbb{R}^d)} \left(1 + \left|\log(\|n\|_{L^p(\mathbb{R}^d)}/\|n\|_{L^1(\mathbb{R}^d)})\right|\right)^{(d-1)/d}.$$

Proof. Let $\alpha > 0$ be a parameter to be chosen later; we decompose the convolution kernel as

$$\nabla u_S = \lambda_d \, \frac{x \, \mathbf{1}_{\{\alpha<|x|<1\}}}{|x|^d} * n + \lambda_d \, \frac{x \, \mathbf{1}_{\{|x|<\alpha\}}}{|x|^d} * n.$$

Therefore, using the Young inequalities, we arrive at

$$\|\nabla u_S\|_{L^{d/(d-1)}(\mathbb{R}^d)} \leq C\|\frac{\mathbf{1}_{\{\alpha<|x|<1\}}}{|x|^{d-1}}\|_{d/(d-1)} \, \|n\|_{L^1(\mathbb{R}^d)} + \|n\|_{L^p(\mathbb{R}^d)} \, \|\frac{\mathbf{1}_{\{|x|<\alpha\}}}{|x|^{d-1}}\|_{L^q(\mathbb{R}^d)},$$

with $1 + \frac{d-1}{d} = \frac{1}{p} + \frac{1}{q}$. Computing the norms of the truncated convolution kernel, we obtain

$$\|\nabla u_S\|_{L^{d/(d-1)}(\mathbb{R}^d)} \leq C|\log(\alpha)|^{(d-1)/d} \, \|n\|_{L^1(\mathbb{R}^d)} + C\|n\|_{L^p(\mathbb{R}^d)} \, \alpha^{(d-1)(1-q)+1}.$$

Finally, we choose

$$\alpha^{(d-1)(1-q)+1} = \frac{\|n\|_{L^1(\mathbb{R}^d)}}{\|n\|_{L^p(\mathbb{R}^d)}},$$

and Lemma 6.9 is proved. $\square$

6.6 The Krein–Rutman theorem

6.6.1 Krein–Rutman theorem

We give here a simplified version of this theorem which extends in infinite dimension the Perron–Froebenius theorem for matrices. See for instance [187], or [72] Appendix to Chapter VIII, Volume 3 (p.193 and following) or Chapter XXI, Section 3.4. for more general statements.

In a vector space E one defines an order thanks to a cone K,

$$x \geq y \Leftrightarrow x - y \in K, \quad \text{and} \quad x > y \Leftrightarrow x - y \in Int(K).$$

A subset K of E is called a *cone* if

(i) $0 \in K$,

(ii) $x, y \in K \Longrightarrow \lambda x + \mu y \in K \quad \forall \lambda \geq 0, \ \mu \geq 0$,

(iii) $x \in K$ and $-x \in K \Longrightarrow x = 0$.

Additionally,

(iv) the cone K is said to be reproducible if $\forall x \in E$, then $\exists y, z \in K$ such that $x = y - z$,

(v) it is said to be normal if $0 \leq x \leq y \Longrightarrow \|x\| \leq \|y\|$.

Theorem 6.5. *(Krein–Rutman) Let $(E, \| \cdots \|)$ be a Banach space and A a continuous and compact linear mapping from E into itself. We assume that A is strongly positive on a closed cone, i.e.,*

$$x \in K \backslash \{0\} \Longrightarrow A(x) > 0.$$

The spectral radius of A, $\rho(A)$ a positive simple eigenvalue of A associated with a positive eigenvector, i.e., $x^0 \in Int(K)$ and it is the only nonnegative eigenvector.

We also recall that for a compact operator, if μ belongs to the spectrum of T, i.e., $T - \mu I$ is not one-to-one, then either $\mu = 0$ or μ is an eigenvalue. See [41], Section 6.2.

This version of the Krein–Rutman theorem does not apply to the L^p spaces because the interior of the positive cone is empty. Therefore it is better to use the space of continuous functions. Then the interior of the nonnegative cone is the set of positive functions. Even though this space is not adapted to PDEs in general, it allows us to solve a 'regularized problem' and it remains to pass to the limit to solve more realistic problems. We illustrate this on the cell division equation.

For applications to PDEs, this first eigenvalue can also be characterized by min-max formulas that extend the usual one in finite dimension (Collatz–Wielandt formula)

$$\rho(A) = \inf\{r > 0 \quad s.t. \quad \exists x \in \mathbb{R}^d \backslash \{0\}, \ x \geq 0, \ A.x \leq rx\}.$$

We refer to [174] for a similar min-max formula for cell division equations and applications to fitness optimization.

6.6.2   Application to cell division equation

As an example we consider a stationary cell division equation set on a bounded interval $[0, R]$ (because the Krein–Rutman theorem requires compactness), for continuous coefficients b and B (as motivated above), and with a boundary condition allowing for positivity:

$$\begin{cases} \frac{\partial}{\partial x}n(x) + \big(\mu + B(x)\big)n(x) - \int_0^R b(x,y)n(y)dy = f(x), \quad 0 \le x \le R, \\[2mm] n(x = 0) = \varepsilon \int_0^R n(y)dy. \end{cases} \tag{6.36}$$

We define the Banach space $E = C^0([0, R])$ (with the sup norm) and assume, that

$$0 \le B(\cdot) \in E, \tag{6.37}$$

$$0 \le b(\cdot, \cdot) \in C^0([0, R] \times [0, R]). \tag{6.38}$$

We prove the following preliminary result that is (very much) improved in Section 4.2.3:

Theorem 6.6. *Assume* (6.37)–(6.38) *and* $\varepsilon > 0$ *small enough, then there is a unique* $\lambda_0 \in \mathbb{R}$ *and* $N, \phi \in C^1([0, R])$, *solutions to the truncated cell division equation and its adjoint,*

$$\begin{cases} \frac{\partial}{\partial x}N(x) + \big(\lambda_0 + B(x)\big)N(x) - \int_0^R b(x,y)N(y)dy = 0, \quad 0 \le x \le R, \\[2mm] N(x = 0) = \varepsilon \int_0^R N(y)dy, \qquad N(x) > 0 \qquad \int_0^R N(x)dx = 1; \end{cases} \tag{6.39}$$

$$\begin{cases} -\frac{\partial}{\partial x}\phi(x) + \big(\lambda_0 + B(x)\big)\phi(x) - \int_0^R b(y,x)\phi(y)dy = \varepsilon\phi(0), \quad 0 \le x \le R, \\[2mm] \phi(x = R) = 0, \qquad \phi(x) \ge 0 \qquad \int_0^R \phi(x)N(x)dx = 1. \end{cases} \tag{6.40}$$

This is a direct application of the Krein–Rutman theorem because one can define the linear operator on E, $A : f \mapsto n$ (solution to (6.36)), for which we prove below that the assumptions of Theorem 6.5 apply. The C^1 regularity is nothing but a consequence of the continuity of B, b and N in (6.39).

Therefore it remains to prove the

Theorem 6.7. *Assume* $\varepsilon > 0$ *small enough and* $\mu > 0$ *large enough. For* $f \in E$, *there is a unique solution* $n \in E$, *to* (6.36). *This defines a continuous linear mapping on* E, $n = A(f)$ *and* A *is compact and strongly positive.*

Proof. First step: construction of A. Fix $f \in E$ and for $m \in E$, we define $n = T(m) \in E$ as the (explicit) solution to

$$\begin{cases} \frac{\partial}{\partial x} n(x) + (\mu + B(x)) n(x) = \int_0^R b(y, x) m(y) dy + f(x), & 0 \leq x \leq R, \\ \\ n(x = 0) = \varepsilon \int_0^R m(y) dy. \end{cases}$$

We prove that T is a strict contraction. Therefore it has a unique fixed point thanks to the Banach–Picard Theorem. This fixed point is a solution to (6.36).

In order to prove that T is a strict contraction, we consider m_1 and m_2 two functions in E, we compute for $n = n_1 - n_2$, $m = m_1 - m_2$,

$$\begin{cases} \frac{\partial}{\partial x} n(x) + (\mu + B(x)) n(x) = \int_0^R b(x, y) m(y) dy, & 0 \leq x \leq R, \\ \\ n(x = 0) = \varepsilon \int_0^R m(y) dy, \end{cases}$$

therefore

$$\begin{cases} \frac{\partial}{\partial x} |n(x)| + (\mu + B(x)) |n(x)| \leq \int_0^R b(x, y) |m(y)| dy, & 0 \leq x \leq R, \\ \\ |n(x = 0)| \leq \varepsilon \int_0^R |m(y)| dy. \end{cases}$$

After integration, we obtain

$$|n(x)| e^{\int_0^x (\mu + B)} \leq \varepsilon \int_0^R |m(y)| dy + \int_0^x e^{\int_0^{x'} (\mu + B)} \int_0^R b(x', y) |m(y)| dy dx'$$

and thus

$$\begin{aligned} |n(x)| &\leq \varepsilon \int_0^R |m(y)| dy + \int_0^x e^{-\int_{x'}^x (\mu + B)} \int_0^R b(x', y) |m(y)| dy dx' \\ &\leq \|m\|_E \left[\varepsilon R + \int_0^x e^{-\int_{x'}^x (\mu + B)} \int_0^R b(x', y) dy dx' \right] \\ &\leq \|m\|_E \left[\varepsilon R + \left\| \int_0^R b(\cdot, y) dy \right\|_{L^\infty} \int_0^x e^{-\mu(x - x')} dx' \right] \\ &\leq \|m\|_E \underbrace{\left[\varepsilon R + \mu^{-1} \left\| \int_0^R b(\cdot, y) dy \right\|_{L^\infty} \right]}_{:=k}. \end{aligned}$$

We can choose, as announced, ε small and μ large so that $k < 1$ and we obtain

$$\|n\|_E \leq k \|m\|_E.$$

Thus T is a strict contraction and we have proved the existence of a solution to (6.36).

Second step: A is continuous. This relies on a general argument which in fact shows that the linear mapping A is Lipschitz continuous. Indeed, arguing as above

$$\begin{aligned} |n(x)| e^{\int_0^x (\mu + B)} &\leq \varepsilon \int_0^R |n(y)| dy \\ &+ \int_0^x e^{\int_0^{x'} (\mu + B)} \int_0^R b(x', y) |n(y)| dy dx' + \int_0^x e^{\int_0^{x'} (\mu + B)} |f(x')| dx'. \end{aligned}$$

and thus
$$|n(x)| \leq k\|n\|_E + \int_0^x |f(x')|dx' \leq k\|n\|_E + \|f\|_E.$$

This indeed proves that
$$\|n\|_E \leq \frac{1}{1-k}\|f\|_E.$$

Third step: A is strongly positive. For $f \geq 0$, the operator T of the first step maps $m \geq 0$ to $n \geq 0$. Therefore the fixed point n is nonnegative. In other words $n = A(f) \geq 0$. If additionally f does not vanish, then n does not vanish either. Therefore $n(0) = \varepsilon \int_0^R n(y)dy > 0$ and thus

$$n(x) \geq n(0) + e^{-\int_0^x (\mu+B)} \int_0^x e^{\int_0^{x'} (\mu+B)} f(x')dx' > 0.$$

Fourth step: A is compact. For $\|f\|_E \leq 1$, the third step proves that n is bounded in E and thus

$$\tfrac{\partial}{\partial x} n = -(\mu + B)n + \int b(x,y)n(y)dy + f$$

is also bounded in E. Therefore by the Ascoli-Arzela theorem the family n is relatively compact in E.   □

Bibliography

[1] S.M. Allen and J.W. Cahn, A macroscopic theory for antiphase boundary motion and its application to antiphase domain coarsening. Acta Metal. **270** (1979), 1085–1095.

[2] M. Adimy, F. Crauste and S. Ruan, A mathematical study of the hematoiesis process with applications to chronic myelogeneous leukemia. SIAM J. Appl. Math. **65** (2005), no. 4, 1328–1352.

[3] M. Adimy and F. Crauste, A nonlinear cellular proliferation model: cells dying out and invariance. C. R. Acad. Sc. Paris I **336** (2003), 559–564.

[4] W.C. Allee, *Animal aggregations: a study in general sociology*. Univ. Chicago Press, Chicago, Illinois, 1931.

[5] W. Alt, Biased random walk models for chemotaxis and related diffusion approximations. J. Math. Biol. **9** (1980), 147–177.

[6] W. Alt and G. Hoffmann, *Biological motion*. L. N. in Biomathematics 89. Springer-Verlag, (1989).

[7] L. Ambrosio, Transport equation and Cauchy problem for BV vector fields.

[8] L. Ambrosio, N. Fusco and D. Pallara, *Functions of Bounded Variations and free discontinuity problems*. Oxford Univ. Press, 2000.

[9] O. Arino, A survey of structured cell population dynamics. Acta Biotheor. **43**, 3–25.

[10] O. Arino, E. Sànchez and G.F. Webb, Necessary and sufficient conditions for asynchronous exponential growth in age structured cell population with quiescence. J. Math. Anal. and Appl. **215** (1997), 499–513.

[11] V.I. Arnold, *Ordinary differential equations*. M.I.T. press, Cambridge, 1990.

[12] A. Arnold, P. Markowich, G. Toscani and A. Unterreiter, On logarithmic Sobolev inequalities, Csiszar-Kullback inequalities and the rate of convergence to equilibrium for Fokker-Planck type equations. Comm. Partial Differential Equations **26** (2001), no. 1-2, 43–100.

[13] F. Baccelli, D.R. McDonald and J. Reynier, A mean field model for multiple TCP connections through a buffer implementing RED. Performance Evaluation **11** (2002), 77–97.

[14] M. Bardi and I. Capuzzo Dolcetta, *Optimal control and viscosity solutions of Hamilton–Jacobi-Bellman equations.* Birkhäuser, Boston, 1997.

[15] C. Bardos, R. Santos and R. Sentis, Diffusion approximation and computation of the critical size. Trans. Amer. Math. Soc. **284** (1984), no. 2, 617–649.

[16] G. Barles, *Solutions de viscosité et équations de Hamilton–Jacobi.* Collec. SMAI, Springer-Verlag, Paris, 2002.

[17] G. Barles L.C. Evans and P.E. Souganidis, Wavefront propagation for reaction diffusion systems of PDE. Duke Math. J. **61** (1990), 835–858.

[18] G. Barles and B. Perthame, Concentrations and constrained Hamilton–Jacobi equations arising in adaptive dynamics. Contemp. Math. To appear.

[19] C. Basdevant, J. Clairambault and F. Lévi, Optimal drup infusion strategies for cancer chronotherapy. ESIAM: Math. Model. Num. Anal. **39** (2005), no. 6, 1069–1086.

[20] B. Basse, B.C. Baguley, E.S. Marshall, W.R. Joseph, B. van Brunt, G. Wake and D.J.N. Wall, A mathematical model for analysis of the cell cycle in cell lines derived from human tumors. J. Math. Biol. **47** (2003), no. 4, 295–312.

[21] W. Beckner, Sharp Sobolev inequalities on the sphere and the Moser-Trudinger inequality. Ann. of Math. **1** (1993), no. 138, 213–242.

[22] F. Bekkal Brikci, J. Clairambault and B. Perthame, A cell population model with proliferation and quiescence structured by molecular content. In preparation.

[23] BenJacob, personal communication (2002).

[24] E. Benoît and M.-J. Rochet, A continuous model of biomass size spectra governed by predation and the effect of fishing on them. J. Th. Biology **226** (2004), 9–21.

[25] H. Berestycki and F. Hamel *Reaction-Diffusion Equations and Propagation Phenomena.* Series Appl. Math. Sci., Springer, 2005.

[26] D. Bernouilli, Essai d'une nouvelle analyse de la mortalité causée par la petite vérole et des avantages de l'inoculation pour la prévention. Mem. Math. Phys. Acad. Roy., **33** (1760), 303–314.

[27] J. Bertoin, On small masses in self-similar fragmentations. Stochatic Process. Appl. **109** (2004), 13–22.

[28] J. Bestel, F. Clément and M. Sorine, A biomechanical model of muscle contraction. In *Medical image computing and computer-assisted intervention*

(MICCAI'01), Lecture Notes in Computer Sc., Vol. 2208, Eds. W.J. Niessen and M.A. Viergever, Springer, 2001.

[29] M.-D. Betterton and M.P. Brenner, Collapsing bacterial cylinders. Phys. Rev. E, **64** (2001), 061904.

[30] D. Beysens, X. Campi and E. Pefferkorn, *Fragmentation phenomena.* World Scientific, Singapore, 1995.

[31] P. Biler, Local and global solvability of some parabolic systems modeling chemotaxis. Adv. Math. Sci. Appl. **8** (1998), no. 2, 715–743.

[32] P. Biler, M. Cannone, I.A. Guerra and G. Karch, Global regular and singular solutions for a model of gravitating particles. Math. Ann. **330** (2004), no. 4, 693–708.

[33] P. Biler, G. Karch, G., P. Laurençot and T. Nadzieja, The 8π problem for radially symmetric solutions of a chemotaxis model in a disc. Topol. Methods Nonlinear Anal. **27** (2006), 133–144.

[34] P. Biler, G. Karch, P. Laurençot and T. Nadzieja, The 8π problem for radially symmetric solutions of a chemotaxis model in the plane. Math. Meth. Appl. Sc. **29** (2006), 1563–1583.

[35] P. Biler and T. Nadzieja, Existence and nonexistence of solutions for a model of gravitational interaction of particles, I. Colloquium Math. **66** (1994), no. 2, 319–334.

[36] A. Blanchet, J. Dolbeault and B. Perthame, Two-dimensional Keller–Segel model: Optimal critical mass and qualitative properties of the solutions. Electron. J. Diff. Eqns. **2006** (2006), no. 44, 1–32.

[37] A. Bonami, D. Hilhorst, E. Logak and M. Mimura, Singular limit of a chemotaxis-growth model. Adv. Differential Equations **6** (2001), no. 10, 1173–1218.

[38] N. Bournaveas, C. Calvez, S. Gutiérrez and B. Perthame, Global existence for a kinetic model of chemotaxis via dispersion and Strichartz estimates. Preprint 2006.

[39] M.P. Brenner, P. Constantin, L.P. Kadanoff, A. Schenkel and S.C. Venkataramani, Diffusion, attraction and collapse. Nonlinearity **12** (1999), no. 4, 1071–1098.

[40] M.P. Brenner, L.S. Levitov and E.O. Budrene, Physical mechanisms for chemotactic pattern formation by bacteria. Biophysical Journal **74** (1998), 1677–1693.

[41] H. Brezis, *Functional analysis.* Masson, 1983.

[42] M. Burger, M. Di Francesco and Y. Dolak, The Keller–Segel model for chemotaxis with prevention of overcrowding: linear vs nonlinear diffusion. Preprint (2005).

[43] R. Bürger, *The mathematical theory of selection, recombination and mutation.* Wiley, 2000.

[44] E. Caglioti, P.-L. Lions, C. Marchioro and M. Pulvirenti, A special class of stationary flows for two-dimensional Euler equations: a statistical mechanics description. Comm. Math. Phys. **143** (1992), 501–525.

[45] À. Calcina and S. Cuadrado, Small mutation rate and evolutionarily stable strategies in infinite dimensional adaptive dynamics. J. Math. Biol. **48** (2004), 135–159.

[46] À. Calcina and J. Soldaña, A model of physiologically structured population dynamics with a nonlinear individual growth rate. J. Math. Biol. **33** (1995), 335–364.

[47] V. Calvez and J. Carrillo, A. Volume effects in the Keller–Segel model: energy estimates preventing blow-up. J. Math. Pures et Appl. **86** (2006), no. 2, 155–175.

[48] V. Calvez and B. Perthame, A Lyapunov function for a two-chemical version of the chemotaxis model. B.I.T. Num. Math., submitted.

[49] V. Calvez, B. Perthame and M. Sharifi tabar, Modified Keller–Segel system and critical mass for the log interaction kernel. Preprint, 2006.

[50] E. Carlen and M. Loss, Competing symmetries, the logarithmic HLS inequality and Onofri's inequality on S^n. Geom. Funct. Anal. **1** (1992), no. 2, 90–104.

[51] J. Carrillo, S. Cuadrado and B. Perthame, Adaptive dynamics via Hamilton–Jacobi approach and entropy methods for a juvenile-adult model. Preprint, 2006.

[52] J. Carrillo, A. Jungel, P. Markowich, G. Toscani and A. Unterreiter, Entropy dissipation methods for degenerate parabolic problems and generalized Sobolev inequalities. Monatsh. Math. **133** (2001), no. 1, 1–82.

[53] C. Cercignani, R. Illner and M. Pulvirenti, *The mathematical theory of dilute gases*, Applied Math. Sciences **106**, Springer, Berlin, 1994.

[54] F. Chalub, P. Markowich, B. Perthame and C. Schmeiser, Kinetic Models for Chemotaxis and their Drift-Diffusion Limits. Monatshefte fuer Mathematik **142** (2004), no. 1–2, 123–141.

[55] F. Chalub and J.-F. Rodrigues, A class of kinetic models for chemotaxis with threshold to prevent overcrowding. Portugaliae Mathematica **63** (2006), no. 2, 1–24.

[56] N. Champagnat, R. Ferrière and S. Méléard, Unifying evolutionary dynamics: From individual stochastic processes to macroscopic models. Theor Popul Biol. **69** (2006), no. 3, 297–321.

[57] D. Chapelle, F. Clément, F. Génot, P. Le Tallec, M. Sorine and J. Urquiza, *A physiologically based model for the active cardiac muscle contraction.* Lecture Notes in Computer Science, Vol. 2230, Eds. T. Katila, I.E. Magnin, P. Clarysse, J. Montagnat and J. Nenonen, Springer, 2002.

[58] M.A.J. Chaplain, A vascular growth, angiogenesis and vascular growth in solid tumors : the mathematical modeling of the stages of tumor development. Math. Comput. modeling , **23** (1996), 47–87.

[59] S. Childress, Chemotactic collapse in two dimensions. Lecture Notes in Biomathematics 55, 61–68, Springer, 1984.

[60] G. Chiorino, J.A.J. Metz, D. Tomasoni and P. Ubezio, Desynchronization rate in cell populations: mathematical modeling and experimental data. J. Theor. Biol. **208** (2001), 185–199.

[61] M. Chipot, S. Hastings and D. Kinderlehrer, Transport in a molecular motor system. M2AN Math. Model. Numer. Anal. **38** (2004), no. 6, 1011–1034.

[62] M. Chipot, D. Kinderlehrer and M. Kowalczyk, A variational principle for molecular motors. Dedicated to Piero Villaggio on the occasion of his 70th birthday. Meccanica **38** (2003), no. 5, 505–518.

[63] J. Clairambault, Ph. Michel and B. Perthame, Circadian rhythm and tumor growth. C. R. Acad. Sc. Paris, Ser. I t. **342** (2006), 17–22.

[64] C. Cocozza-Thivent and R. Eymard, Approximation of the marginal distributions of a semi-Markov process using a finite volume scheme. ESAIM: M2AN **38** (2004), no. 5, 853–876.

[65] E.A. Coddington and N. Levinson, *Theory of ordinary differential equations.* New York, McGraw-Hill, 1955.

[66] L. Corrias and B. Perthame, Critical space for the parabolic-parabolic Keller–Segel model in $\mathbb{R}^d$. C. R. Acad. Sc. Paris, Ser. I **342** (2006), 745–750.

[67] L. Corrias, B. Perthame and H. Zaag, A chemotaxis model motivated by angiogenesis. C. R. Acad. Sci. Paris, Ser. I **336** (2003), 141–146.

[68] L. Corrias, B. Perthame and H. Zaag, A model motivated by angiogenesis. Milan J. of Math. **72** (2004), 1–29.

[69] M.G. Crandall and P.-L. Lions, User's guide to viscosity solutions of second order partial differential equations. Bull. Amer. Math. Soc. **27** (1992), 1–67.

[70] J.M. Cushing, *An Introduction to Structured Population Dynamics.* CBMS-NSF, Regional conference series in applied mathematics, SIAM, 1998.

[71] C.M. Dafermos, *Hyperbolic conservation laws in continuum physics*. GM 325, Springer, 1999.

[72] R. Dautray and J.-L. Lions, *Mathematical analysis and numerical methods for sciences and technology*. Springer, 1990.

[73] E. De Angelis and P.-E. Jabin, Qualitative analysis of a mean field modeling of tumor-immune system competition. Math. Models Methods Appl. Sci. **13** (2003), 187–206.

[74] E. De Angelis and L. Mesin, On the kinetic (cellular) theory. Conceptual framework on modeling the immune response. Math. Models Methods Appl. Sci. **11** (2001), no. 9, 1609–1630.

[75] P. Degond, T. Goudon and F. Poupaud, Diffusion limit for nonhomogeneous and non-micro-reversible processes. Indiana Univ. Math. J. **49** (2000), no. 3, 1175–1198.

[76] M. Del Pino and J. Dolbeault, The optimal euclidian L^p Sobolev logarithmic inequality. J. Funct. Anal. **197** (2003), no. 1, 151–161.

[77] L. Derbel, Analysis of a new model for tumor-immune system comptition including long time scale effects. ENS report DMA-04-09.

[78] L. Derbel and P.-E. Jabin, The set of concentration for some hyperbolic models of chemotaxis. Preprint, 2006.

[79] F. Dercole and S. Rinaldi, Evolution of cannibalistic traits: scenarios derived from adaptive dynamics. Th. Pop. Biology **62** (2002), 365–374.

[80] U. Dieckmann and R. Law, The dynamical theory of coevolution: A derivation from a stochastic ecological processes. J. Math. Biol. **34** (1996), 579–612.

[81] O. Diekmann, A beginner's guide to adaptive dynamics. In: *Mathematical modeling of population dynamics*, ed. R. Rudnicki, Banach Center Publications, Vol. 63 (2004), 47–86.

[82] O. Diekmann and J.A.P. Heesterbeck, *Mathematical Epidemiology of Infectious Diseases*. Wiley, 2000.

[83] O. Diekmann, P.-E. Jabin, S. Mischler and B. Perthame, The dynamics of adaptation : an illuminating example and a Hamilton–Jacobi approach. Th. Pop. Biology, **67** (2005), no. 4, 257–271.

[84] R.J. DiPerna and P.-L. Lions, Ordinary differential equations, transport theory and Sobolev spaces. Invent. Math. **98** (1989), no. 3, 511–547.

[85] Y. Dolak and C. Schmeiser, A kinetic theory approach for resolving the chemotactic wave paradox. Mathematical modeling & computing in biology and medicine, 171–177, Milan Res. Cent. Ind. Appl. Math. MIRIAM Proj., 1, Esculapio, Bologna, 2003.

[86] Y. Dolak and C. Schmeiser, The Keller–Segel model with logistic sensitivity function and small diffusivity. SIAM J. Appl. Math. **66** (2005), no. 1, 286–308 (electronic).

[87] J. Dolbeault, D. Kinderlehrer and M. Kowalczyk, Remarks about the flashing rachet. Partial differential equations and inverse problems, 167–175, Contemp. Math., 362, Amer. Math. Soc., Providence, RI, 2004.

[88] J. Dolbeault and B. Perthame, Optimal critical mass in the two dimensional Keller–Segel model in $\mathbb{R}^2$. C. R. Math. Acad. Sci. Paris **339** (2004), no. 9, 611–616.

[89] C. Doss-Bachelet, J.-P. Françoise and C. Piquet, Bursting oscillations in two coupled Fitzhugh–Nagumo systems. ComPlexUs **2** (2003), 101–111.

[90] R. Durrett, Crabgrass, measles, and gypsy moths: an introduction to modern probability. Bull. Amer. Math. Soc. (N.S.) **18** (1988), no. 2, 117–143.

[91] R. Durrett, *Ten lectures on particle systems*. Lectures on probability theory (Saint-Flour, 1993), 97–201, Lecture Notes in Math. 1608, Springer, 1995.

[92] J. Dyson, R. Villella-Bressan and G. Webb, A nonlinear age and maturity structured model of population dynamics, II. Chaos. J. Math. Anal. Appl. **242** (2000), No. 2, 255–270.

[93] N. Echenim, D. Monniaux, M. Sorine and F. Clément, Multi-scale modeling of the follicle selection process in the ovary. Math. Biosci. **198** (2005), no. 1, 57–79.

[94] N. Echenim, M. Sorine and F. Clément, A multiscale model for the controlled selection process of ovulary follicles. Preprint, 2005.

[95] L. Edelstein-Keshet, *Mathematical models in biology*, 2nd edition, 2005.

[96] R. Erban and Hyung Ju Hwang, Global existence results for complex hyperbolic models of bacterial chemotaxis. Preprint, 2006 (arXiv:math.AP/0602139).

[97] S.E. Esipov and J.A. Shapiro, Kinetic model of Proteus Mirabilis swarm colony development. J. Math. Biology **36** (1998), 249–268.

[98] L.C. Evans, *Partial Differential Equations*. Graduate Studies in Mathematics Vol. 19, American Mathematical Society, 1998.

[99] L.C. Evans, A survey of entropy methods for partial differential equations. Bull. Amer. Math. Soc. **41** (2004), 409–438.

[100] W. Feller, *An introduction to probability theory and applications*. Wiley, New-York, 1966.

[101] R. Ferrière, S. Méléard, N. Fournier and N. Champagnat, The mathematical of Darwinian evolution: from stochastic individual processes to adaptative dynamics. Preprint, 2004.

[102] P.C. Fife, *Mathematical aspects of reacting and diffusing systems.* L. N. in biomathematics 28, Springer, 1979.

[103] F. Filbet, P. Laurencot and B. Perthame, Derivation of Hyperbolic Models for Chemosensitive Movement. J. Math. Biology. **50** (2004), no. 2, 189–207.

[104] F. Filbet and C.-W. Shu, Approximation of Hyperbolic Models for Chemosensitive Movement. SIAM J. Sci. Comput. vol. **27** (2005), 850–872.

[105] R.A. Fisher, *The genetical theory of natural selection.* Clarendon Press, 1930. 2nd ed.: Dover, 1958. 3rd ed., présentée et annotée par Henry Bennett: Oxford Univ. Press, 1999.

[106] R. Fitzhugh, Impulses and physiological states in theoretical models of nerve membrane. Biophys. J. **1** (1961), 445–466.

[107] M.A. Fontelos, A. Friedman and B. Hu., Mathematical analysis of a model for the initiation of angiogenesis. SIAM J. Math. Anal. **33** (2002), 1330–1355.

[108] D. Fouchet, M. Langlais and D. Pontier, University Cl. Bernard (Lyon), Technical report, 2002.

[109] J.-P. Françoise, *Oscillations en biologie.* Coll. Math. et Appl., SMAI, Springer, Paris, 2005.

[110] M. Freidlin, *Functional integration and partial differential equations.* Annals of Mathematics Studies **109**, Princeton University Press, Princeton, NJ, 1985.

[111] P. Friedl and K. Wolf, Tumour-cell invasion and migration: diversity and escape mechanisms. Nature reviews (cancer) **3** (2003), 362–374.

[112] H. Frid, P.-E. Jabin and B. Perthame, Global Stability of Steady Solutions in Virus Dynamics. ESAIM:M2AN, Vol. 37, No.4, 2003.

[113] A. Gamba, D. Ambrosi, A. Coniglio, A. de Candia, S. Di Talia, E. Giraudo, G. Serini, L. Preziosi and F. Bussolino, Percolation, morphogenesis, and Burgers dynamics in blood vessels formation. Phys. Rev. Lett. **90** (2003), 118101.

[114] H. Gajewski and K. Zacharias, Global behavior of a reaction diffusion system modeling chemotaxis. Math. Nachr. **195** (1998), 77–114.

[115] S.A.H. Geritz, E. Kisdi, G. Meszena and J.A.J. Metz, Evolutionary singular strategies and the adaptive growth and branching of the evolutionary tree. Evolutionary Ecology **12** (1998), 35–57.

[116] S. Ghosal and S. Mandre, A simple model illustrating the role of turbulent life on phytoplankton blooms. J. Math. Biol. **46** (2003), no. 4, 333–346.

[117] S. Ghosal, M. Rogers and A. Wray, The turbulent life of phytoplankton. Center for turbulence research, proceedings of the summer program, 2000.

[118] D. Gilbarg and N.S. Trudinger, *Elliptic partial differential equations of second order*. Springer, 1983.

[119] R.T. Glassey, *The Cauchy problem in kinetic theory*. SIAM publications, Philadelphia, 1996.

[120] A. Goldbeter, *Biochemical oscillators and cellular rhythms*. Cambridge University Press, 1997.

[121] I. Golding, Y. Kozlovsky, I. Cohen and E. Ben-Jacob, Studies of bacterial branching growth using reaction-diffusion models for colonial development. Physica A **260** (1998), 510–554.

[122] Th. Goudon, Hydrodynamic limit for the Vlasov-Poisson-Fokker-Planck system: analysis of the two dimensional case. Math. Models Methods Appl. Sci. **15** (2005), no. 5, 737–752.

[123] E. Grenier, Quelques modèles en médecine. (French) [Some models in medicine] Journées "Équations aux Dérivées Partielles", Exp. No. VI, 23 pp., École Polytech., Palaiseau, 2004.

[124] P. Gwiazda and B. Perthame, Invariants and exponential rate of convergence to steady state in the renewal equation. Markov Processes and Related Fields (MPRF) **2** (2006), 413–424.

[125] M.A. Henson, Dynamic modeling of microbial cell populations. Current opinion in biotechnology **14** (2003), 460–467.

[126] M.A. Herrero and J.J.L. Velázquez, Singularity patterns in a chemotaxis model. Math. Annal. **306** (1996), 583–623.

[127] M.A. Herrero, E. Medina and J.J.L. Velázquez, Finite-time aggregation into a single point in a reaction-diffusion system. Nonlinearity **10** (1997), no. 6,1739–1754.

[128] T. Hillen, *Mesoscopic and macroscopic models for mesenchymal motion*. Course given at Teneriffa, Sept. 2005. J. Math. Biol. To appear.

[129] T. Hillen and H. Othmer, The diffusion limit of transport equations derived from velocity-jump processes. SIAM J. Appl. Math. **61** (2000), no. 3, 751–775.

[130] T. Hillen and K. Painter, Global existence for a parabolic chemotaxis model with prevention of overcrowding. Adv. Appl. Math. **26** (2001), 280–301.

[131] T. Hillen, K. Painter and C. Schmeiser, Global existence for chemotaxis with finite sampling radius. DCDS-B. To appear.

[132] T. Hillen and K. Painter, Volume-filling and quorum-sensing in models for chemosensitive movement. Canadian Appl. Math. Quarterly **10** (2002), no. 4, 501–543.

[133] M.W. Hirsch and S. Smale, *Differential equations, dynamical systems and linear algebra*. Academic Press, 1974.

[134] J. Hofbauer and K. Sigmund, Adaptive dynamics and evolutionary stability. Appl. Math. Lett. **3** (1990), 75–79.

[135] J. Hofbauer and K. Sigmund, *The theory of evolution and dynamical systems*. Cambridge Univesity Press, 1998. And *Evolutionary games and population dynamics*. London Mathematical Society, Student texts 7. Cambridge Univesity Press, 2002.

[136] J. Hofbauer and K. Sigmund, Evolutionary game dynamics. Bull. Am. Math. Soc. **40** (2003), no. 4, 479–519.

[137] D. Horstmann, From 1970 until now: The Keller–Segel model in chemotaxis and its consequences I. Jahresberichte der DMV **105** (2003), 103-165.

[138] Horstmann, D. From 1970 until now: The Keller–Segel model in chemotaxis and its consequences II. Jahresberichte der DMV, **106**, (2004) pp. 51-69.

[139] D. Horstmann and M. Winkler, Boundedness vs. blow-up in a chemotaxis system. J. Diff. Eq. **215** (2005), 52–107.

[140] J. Howard, Mechanics of motor proteins and the cytoskeleton. Sinauer Ass., Inc., 2001.

[141] A.F. Huxley, Muscle structure and theories of contractions. In *Progress in biophysics and biological chemistry*, Vol. 7, Pergamon Press, 1987.

[142] H.J. Hwang, K. Kang and A. Stevens, Global solutions of nonlinear transport equations for chemosensitive movement. SIAM J. Math. Anal. **36** (2005), no. 4, 1177–1199.

[143] M. Iannelli, *Mathematical Theory of Age-Structured Population Dynamics*. Applied Math. Monographs, CNR, Giardini Editori e Stampatori in Pisa, 1995.

[144] H. Ishii and I. Takaji, Global stability of stationary solutions to a nonlinear diffusion equation in phytoplankton dynamics. J. Math. Biology **16** (1982), 1–24.

[145] W. Jäger and S. Luckhaus, On explosions of solutions to a system of partial differential equations modeling chemotaxis. Trans. Amer. Math. Soc. **239** (1992), no. 2, 819–821.

[146] F. Jülicher, A. Ajdari and J. Prost, Modeling molecular motors. Rev. Modern Phys. Vol. **69** (1997), no. 4, 1269–1281.

[147] D. Kaiser, Personal communication, 2003.

[148] E.F. Keller and L.A. Segel, Initiation of slide mold aggregation viewed as an instability. J. Theor. Biol. **26** (1970), 399–415.

[149] E.F. Keller and L.A. Segel, Model for chemotaxis. J. Theor. Biol. **30** (1971), 225–234.

[150] Keener, J. and Sneyd, J. *Mathematical physiology.* Interdisciplinary Applied Mathematics, 8. Springer-Verlag, New York, 1998.

[151] W.O. Kermack and A.G. McKendrick, A contribution to the mathematical theory of epidemics. Part I. Proc. Roy. Soc. A **115** (1927), 700–721. Part II. Proc. Roy. Soc. A **138** (1932), 55–83.

[152] D. Kinderlehrer and M. Kowalczyk, Diffusion-mediated transport and the flashing ratchet. Arch. Ration. Mech. Anal. **161** (2002), no. 2, 149–179.

[153] A.N. Kolmogorov, I.G. Petrovski and N.S. Piskunov, Étude de l'équation de la diffusion avec croissance de la quantité de matière et son application à un problème biologique. Bulletin Université d'État de Moscou (Bjul. Moskowskogo Gos. Univ.), Série internationale **A1** (1937), 1–26.

[154] R. Kowalczyk, Preventing blow-up in a chemotaxis model. J. Math. Anal. Appl. **305** (2005), no. 2, 566–588.

[155] M. Ledoux, *The concentration of measure phenomenon.* AMS Math. Surveys and Monographs 89, Providence, 2001.

[156] H.A. Levine, M. Nilsen-Hamilton and B.D. Sleeman, Mathematical modeling of the onset of capillary formation initiating angiogenesis. J. Math. Biol. **42** (2001), 195–238.

[157] E.H. Lieb and M. Loss, *Analysis.* 2nd ed, AMS Graduate Studies in Math., 2001.

[158] P.-L. Lions, *Mathematical topics in fluid mechanics, incompressible models.* Oxford Lecture Series in Mathematics and its Applications Vol. 1. Oxford University Press, 1996.

[159] C. Lobry and J. Harmand, A new hypothesis to explain the coexistence of n species in the presence of a single resource. C. R. Biologies **329** (2006), 40–46.

[160] C. Lobry, F. Mazenc and A. Rapaport, Persistence in ecological models of competition for a single resource. C. R. Math. Acad. Sci. Paris **340** (2005), no. 3, 199–204.

[161] E.D. McGrady and R.M. Ziff, Shattering transition in fragmentation. Phys. Rev. Letters **58** (1987), no. 9, 892–895.

[162] A.G. McKendrick, Applications of mathematics to medical problems, Proc. Edinburgh Math. Soc. **44** (1926), 98–130.

[163] M.C. Mackey and A. Rey, Multistability and boundary layer development in a transport equation with retarded arguments. Can. Appl. Math. Quart. **1** (1993), 1–21.

[164] P.K. Maini, Applications of mathematical modeling to biological pattern formation. In: *Coherent structures in complex systems* (Sitges, 2000), 205–217, Lecture Notes in Phys. 567, Springer, Berlin, 2001.

[165] T.R. Malthus, *An essay on the principle of population.* 1st ed., London, 1798.

[166] N.V. Mantzaris, S. Webb and H.G. Othmer, Mathematical modeling of tumor-induced angiogenesis. J. Math. Biol. **49** (2004), no. 2, 111–187.

[167] P.A: Markowich, C.A. Ringhofer and C. Schmeiser, *Semiconductor equations.* Springer-Verlag, New York, 1989.

[168] A. Marrocco, 2D simulation of chemotactic bacteria agreggation. ESAIM:M2AN **37** (2003), no. 4, 617–630.

[169] J. Maynard Smith, *Evolution and the theory of games.* Cambridge Univ. Press, 1982.

[170] J. Maynard-Smith and G.R. Price, The logic of animal conflict. Nature **246** (1973), 15–18.

[171] J.A.J. Metz and O. Diekmann, *The dynamics of physiologically structured populations.* L.N. in biomathematics 68, Springer, 1986.

[172] J.A.J. Metz, S.A.H. Geritz, G. Meszena, F.J.A. Jacobs and J.S. van Heerwaarden, Adaptive dynamics, a geometrical study of the consequences of nearly faithful reproduction. In: *Stochastic and spatial structures of dynamical systems* (Amsterdam, 1995), 183–231, Konink. Nederl. Akad. Wetensch. Verh. Afd. Natuurk. Eerste Reeks, 45, North-Holland, Amsterdam, 1996.

[173] P. Michel, Existence of a solution to the cell division eigenproblem. Model. Math. Meth. Appl. Sci. **16**, suppl. issue 1 (2006), 1125–1153.

[174] P. Michel, Optimal proliferation rate in a cell division model. Preprint, 2006.

[175] P. Michel, S. Mischler, and B. Perthame, General entropy equations for structured population models and scattering. C.R. Acad. Sc. Paris, Sér. I **338** (2004), 697–702.

[176] P. Michel, S. Mischler and B. Perthame, General relative entropy inequality: an illustration on growth models. J. Math. Pures et Appl. Vol. **84** (2005), no. 9, 1235–1260.

[177] S. Mischler, B. Perthame and L. Ryzhik, Stability in a Nonlinear Population Maturation Model. Math. Mod. Meth. Appl. Sci. **12** (2002), no. 12, 1751–1772.

[178] J.D. Murray, *Mathematical biology*, Vol. 1 and Vol. 2, 2nd ed., Springer, 2002.

[179] T. Nagai, Blow-up of radially symmetric solutions to a chemotaxis system. Adv in Math. Appl. Sc. **5** (1995), 581–601.

[180] T. Nagai, T. Senba and T. Suzuki, Chemotactic collapse in a parabolic system of mathematical biology. Hiroshima Math. J. **30** (2000), no. 3, 463–497.

[181] T. Nagai, T. Senba and K. Yoshida, Application of the Trudinger-Moser inequality to a parabolic system of chemotaxis. Funk. Ekv. **40** (1997), 411–433.

[182] J. Nagumo, S. Arimoto and S. Yoshizawa, An active pulse transmission line simulating nerve axon. Proc. Inst. Radio Eng. **50** (1962), 2061–2070.

[183] C. Neuhauser, Mathematical challenges in spatial ecology. Notices of AMS **48** (2002), no. 11, 1304–1314.

[184] T.W. Ng, G. Turinici and A. Danchin, A double epidemic model for the SARS propagation. BMC infectious dideases **3** (2003), 19.

[185] J. Nieto, F. Poupaud and J. Soler, High field limit for the Vlasov-Poisson-Fokker-Planck system. Arch. Ration. Mech. Anal. **158** (2001), no. 1, 29–59.

[186] M. Nowak and R. May, *Virus Dynamics: Mathematical Principles of Immunology and Virology*. Oxford University Press, Oxford, 2000.

[187] R.D. Nussbaum, Hilbert's projective metric and iterated nonlinear maps. Mem. Amer. Math. Soc. **75** (1988), no. 39.

[188] A. Okubo and S.A. Levin, *Diffusion and ecological problems, modern perspectives*. Interdisciplinary and Applied Mathematics, Springer, 2001.

[189] H.G. Othmer, S.R. Dunbar and W. Alt, Models of dispersal in biological systems, J. Math. Biol. **26** (1988), 263–298.

[190] H.G. Othmer and T. Hillen, The diffusion limit of transport equations II: chemotaxis equations. SIAM J. Appl Math. **62** (2002), 1222–1250.

[191] H.G. Othmer and A. Stevens, Aggregation, blowup and collapse : the ABC's of taxis in reinforced random walks. SIAM J. Appl. Math. **57** (1997), 1044–1081.

[192] G.C. Papanicolaou, Asymptotic analysis of transport processes. Bull. Amer. Math. Soc. **81** (1975), 330–392.

[193] E. Pate, Mathematical analysis of the generation of force and motion in concentrating muscle. In: *Tutorial in math. biosciences II* (J. Sneyd, ed.), L.N. in Math. 1867, Springer, 2005.

[194] C.S. Patlak, Random walk with persistence and external bias. Bull. Math. Biol. Biophys. **15** (1953), 311–338.

[195] A.S. Perelson and G. Weisbuch, Immunology for physicists. Reviews of modern physics **69** (1997), no. 4, 1219–1267.

[196] B. Perthame, *Kinetic formulation of conservation laws*, Oxford Univ. Press., 2002.

[197] B. Perthame, *Mathematical Tools for Kinetic Equations*. Bull. Amer. Math. Soc. **41** (2004), no. 2, 205–244.

[198] B. Perthame and L. Ryzhyk, Exponential decay for the fragmentation or cell-division equation. J. Diff. Eq. **210** (2005), 155–177.

[199] B. Perthame and P.E. Souganidis, Front propagation for a jump process model arising in spatial ecology. DCDS **13**(2005), no. 5, 1235–1248.

[200] K. Post, A system of nonlinear partial differential equations modeling chemotaxis with sensitivity functions. PdD dissertation, Humboldt-Univ. Berlin, 1999.

[201] M. Primicerio and B. Zaltzman, Global in time solution to the Keller–Segel model of chemotaxis. Preprint. See also: Free boundary in radial symmetric chemotaxis. In: "WASCOM 2001" — 11th Conference on Waves and Stability in Continuous Media (Porto Ercole), 416–427, World Scientific, River Edge, NJ, 2002.

[202] M. Rascle and C. Ziti, Finite time blow-up in some models of chemotaxis. J. Math. Biol. **33** (1995), 388–414.

[203] C.S. Reynolds, *The ecology of freshwater phytoplancton*. Cambridge University Press, 1984.

[204] A.M. de Roos, A gentle introduction to physiologically structured population models. In: *Structured-population models in marine terrestrial, and freshwatersystems* (S. Tuljapurkar and H. Caswell, ed.), Chapman & Hall, Population and Community Biology Series, Vol. 18.

[205] M. Rotenberg, Transport theory for growing cell populations. J. Theor. Biology **103** (1983), 181–199.

[206] E.A. Sausville, Complexities in the development of cyclin-dependent kinase inhibitor drugs. Trends in Molecular Medicine **8** (2002) no. 4 (Suppl.).

[207] H.R. Schwetlick, Travelling fronts for multidimensional nonlinear transport equations. Ann. Inst. H. Poincaré, Analyse non linéaire **17** (2000), no.. 4, 523–550.

[208] T. Senba and T. Suzuki, Chemotactic collapse in parabolic-elliptic systems of mathematical biology. Adv. Differential Equations **6** (2001), 21–50.

[209] G. Serini, D. Ambrosi, E. Giraudo, A. Gamba, L. Preziosi and F. Bussolino, Modeling the early stages of vascular network assembly. The EMBO Journal **22** (2003), 1771–1779.

[210] D. Serre, *Les matrices*. Dunod, Paris, 2001.

[211] H.L. Smith and P. Waltman, *The theory of the chemostat: dynamics of microbial competition.* Cambridge Univ. Press, 1994.

[212] D.R. Soll, Behavioral studies into the mechanism of eukaryotic chemotaxis. J. Chemical Ecology **16** (1990), no. 1, 133–150.

[213] P.E. Souganidis, *Front propagation: theory and applications.* CIME course on 'Viscosity solutions', Lecture Notes in Math., Springer, 1998.

[214] A. Stevens, Derivation of chemotaxis-equations as limit dynamics of moderately interacting stochastic many particle systems. SIAM J. Appl. Math. **61** (2000), 183–212.

[215] Y. Sugiyama and H. Kuni, Global existence and decay properties for a degenerate Keller–Segel model with a power factor in drift term. J. diff. Eq. **227** (2006), no. 1, 333–364.

[216] L. Tartar, Une nouvelle caractérisation des *M* matrices. (French) Rev. Française Informat. Recherche Opérationelle **5** (1971), Ser. R-3, 127–128.

[217] C.H. Taubes, *Modeling lectures on differential equations in biology.* Prentice-Hall, 2001.

[218] H.R. Thieme, *Mathematics in population biology.* Princeton University Press. Princeton, 2003.

[219] A. Triller and D. Choquet, Surface trafficking of receptors between synaptic and extrasynaptic membranes: and yet they do move. Review, Trends in Neurosciences, 2006.

[220] A.M. Turing, The chemical basis of morphogenesis. Phil. Trans. Roy. Soc. London B **237** (1952), 37–72.

[221] J.J. Tyson and O. Diekmann, Sloppy size control of the cell division cycle. J. Theor. Biol. **118** (1986), 405–426.

[222] B. van der Pol, On oscillation hysteresis in a simple triode generator. Phil. Mag. **43** (1922), no. 6, 700–719.

[223] B. van der Pol and J. van der Mark, The heartbeat considered as a relaxation oscillator, and an electrical model of the heart. Phil. Mag. **6** (1928), suppl., 763.

[224] N.G. van Kampen, *Stochastic processes in physics and chemistry.* Lecture Notes in Mathematics 888. North-Holland, Amsterdam, New York, 1981.

[225] J.J.L. Velázquez, Stability of some mechanisms of chemotactic aggregation. SIAM J. Appl. Math. **62** (2002), 1581–1633.

[226] J.J.L. Velázquez, Points dynamics in a singular limit of the Keller–Segel model II. Formation of the concentration region. SIAM J. Appl. Math. **64** (2004), no. 4, 1224–1248.

[227] C. Villani, A survey of mathematical topics in kinetic theory. In: *Handbook of fluid mechanics* (S. Friedlander and D. Serre, eds.), Tome I, chapitre 2. Elsevier, 2002.

[228] C. Villani, *Topics in optimal transportation*. Graduate Studies in Mathematics 58. American Mathematical Society, Providence, RI, 2003.

[229] V. Volterra, Variazioni e fluttuazioni del numero d'individui in specie animali convisenti. Mem. accad. Lincei **2** (1926), no. 6, 31–113.

[230] H. VonFoerster, Some remarks on changing populations. In: *The kinetics of cellular proliferation* (F. Stohlman, ed.), Grune and Strutton, New York, 1959.

[231] G.F. Webb, *Theory of nonlinear age-dependent population dynamics*. Marcel Dekker, New York, 1985.

[232] L.J. White and G.F. Medley, Microparasite population dynamics and continuous immunity. Proc. Roy. Soc. London B **265** (1998), 1977–1983.

Index

CHAPTER 5
Thinking as Redemption: Plotinus between Plato and Augustine

1. Ferdinand Christian Bauer, *Die christliche Gnosis* (Tübingen, 1835). And later, Hans Jonas, *Gnosis und spätantiker Geist* (Göttingen, 1934).
2. In this context it is significant that Proclus's commentary on Plato's *Parmenides* considers only the dialogue's first hypothesis, which was accessible for such interpretation. On this see my "Der platonische *Parmenides* und seine Nachwirkung," in *Gesammelte Werke* (Tübingen, 1991), 7:313–27.
3. Compare my remarks on the concept of the spring in Appendix 5 of *Truth and Method*, trans. Joel Weinsheimer and Donald Marshall (New York, 1989), pp. 501–2.
4. Fortunate first made, Creation's spoiled darlings,
    mountain ranges, peaks reddened by the first of all suns, — pollen
        of a blossoming God,
    pivots of light, corridors, stairways, thrones, spaces formed of sheer
        Being, shields of ecstasy, resounding
    tempests of rapture, and suddenly, every one, *mirrors:* drawing back
        into your own faces
    the same beauty you brilliantly beamed out.
        The Duino Elegies, trans. Leslie Norris and Alan Keele
        (Columbia, SC, 1993), "Second Elegy," p. 9.

CHAPTER 6
Myth in the Age of Science

1. On this topic see the first two of my Hölderlin studies in *Gesammelte Werke*, vol. 9: "Hölderlin und die Antike" and "Hölderlin und das Zukünftige."
2. Christian Hartlich and Walter Sachs, *Der Ursprung des Mythosbegriffes in der modernen Bibelwissenschaft* (Tübingen, 1952).
3. Introduction to *Philosophie der Mythology*, in *Sämtliche Werke*, ed. K. F. A. Schelling (Stuttgart and Augsburg, 1856), 11:193.
4. Ibid., p. 198.
5. On the problem of applying this concept to Greek religion, see my essay "Sokrates' Frömmigkeit des Nichtwissens," in my *Gesammelte Werke*, 7:83–117.
6. Walter F. Otto, *Die Götter Griechenlands* (Bonn, 1929); *Dionysos: Mythos und Kultus* (Frankfurt, 1933).

NOTES

CHAPTER 2
On the Possibility of a Philosophical Ethics

1. Kant, *Grundlegung zur Metaphysik der Siften* (Berlin, 1902), 4:404 ff.
2. Hegel, *Phenomenologie der Geist*, ed. J. Hoffmeister (Leipzig and Hamburg, 1905), pp. 301 ff.
3. G. Krüger, *Philosophie und Moral in der Kantischen Kritik* (Tübingen, 1931; 2d ed., 1967).
4. As the editor of the *Walberger Studien* remarks, Thomas Aquinas emphasizes that "conscience" refers to an act, and only in an extended sense to a habitus underlying it (*Summa theologica* 1:79, 13; vol. 6).
5. Underlying my discussion here is my essay "Praktisches Wissen" of 1930, which has been published for the first time in my *Gesammelte Werke* (Tübingen, 1985–95), 5:230 ff.
6. *Nicomachean Ethics* Z 5, 1140b17: εὐθὺς οὐ φαίνεται ἡ ἀρχή.
7. Ibid., *B* 3, 1105b12 ff.: ἐπὶ τὸν λόγον καταφεύγοντες.
8. See my essay "Freundschaft und Selbsterkenntnis," in the *Festschrift* for U. Höllscher: *Würzburger Abhandlungen zur Altertumswissenschaft* NF 1 (1985): 25–33 (*Gesammelte Werke*, 7:396–406).
9. See my review of Gauthier-Jolif's commentary on the *Nicomachean Ethics*, in *Philosophische Rundschau* 10 (1962): 293 ff. (*Gesammelte Werke*, 6:302–6).
10. More on this in *Truth and Method*, rev. ed., trans. Joel Weinsheimer and Donald G. Marshall (New York, 1992), pp. 318 ff. (*Gesammelte Werke*, 1:324 ff., 2:401 ff.).
11. Plato, *Republic* 302a.

CHAPTER 3
On the Divine in Early Greek Thought

1. W. Jaeger, *Die Theologie der früheren griechischen Denker* (Stuttgart, 1953), p. 43.
2. Karl Reinhardt, *Parmenides* (Bonn, 1916), pp. 251 f.
3. Bruno Snell, *Die Entdeckung des Geistes* (Hamburg, 1946), pp. 205 f.

4. H. Boeder has elaborated on this in *Grund und Gegenwart als Frageziel der frühgriechischen Philosophie* (The Hague, 1962).

5. *Metaphysics* Λ 8, 1074b1 ff.

6. Ibid., *B* 4, 1000a9 ff.

7. The cardinal mistake in Werner Jaeger's brilliant analysis of the "Theologie der frühen Griechen," I think, is that he reverses this relationship. The same is the case for O. Gigon's careful and prudent lecture to be seen in the Fondation Hardt; the appendix especially, in my opinion, addresses a question to the tradition that is inappropriate to it. Does this "theology" of the Ionians and Eleatics have anything at all to do with the factical religion of the time?

8. Karl Joël, *Der Ursprung der Naturphilosophie aus dem Geiste der Mystik* (Jena, 1906).

9. F. M. Cornford, *Principium Sapientiae: The Origins of Greek Philosophical Thought* (Cambridge, 1952).

10. Plato, *Sophist* 248e8.

11. Gadamer, "Vorgestalten der Reflexion," in *Subjectivität und Metaphysik*, *Festschrift* for W. Cramer (Frankfurt am Main, 1966), pp. 128–43 (*Gesammelte Werke* 6:116–28).

12. See H. J. Krämer, *Der Ursprung der Geistmetaphysik* (Amsterdam, 1964), pp. 193 ff. There Krämer admittedly completely exceeds the immanent interpretation of the conversation and tries to answer on new ground a question that is completely unanswerable from it—namely, whether the transcendent world of Ideas or the "totality" is meant by παντελῶς ὄν. This question is irrelevant to the dialogic theme of the Sophist. No more can the philosopher follow the Heracliteans than those who accept the one *or* the many Ideas. Parmenides and a dogmatic Platonism are equally valid! Thus the "dynamic character of Plato's cosmos of ideas" merely interpolates a new dogmatism into a conversation that was trying to answer for the first time the question about what it is that is here called "dynamic"—to become at one with ὄν and μὴ ὄν through knowledge of the structure of the λόγος.

13. *Laws* X, 898d ff. is a certain proof of this, as Jaeger has already seen.

14. *De caelo* B13, 294a30 f.

15. *Fragmente der Vorsokratiker* (hereafter *VS*), (Berlin, 1906) 2, A 2.

16. See Boeder, *Grund und Gegenwart*.

17. *VS* 18 B 8, 4 and 8, 42–9 with *VS* 20 B 2 and b 4.

18. Instead, it follows from the demiurge's intention that he leave nothing out of the four elements that he brings together (*Timaeus* 32c, 33a)—and even that is still criticized as a representation that is much too extrinsic,

because with respect to the σῶμα, the soul is first. It is only with Aristotle, who dissolves the myth of the demiurge and recognizes only the (σύνολον), that the argument from the ὕλη receives full weight.

19. The dominant role that this aspect of the *Timaeus* has played from Xenocrates on, that finally led to shifting the ideas into the divine spi[rit] we can now see thanks to the investigations of P. Merlan, *From Plato[n]ism to Neoplatonism* (The Hague, 1953) and more clearly of H. J. Krä[mer] (*Der Ursprung*). Of course, with the exception of Aristotle, the Acad[emy] translated the Timaeus myth into serious doctrine in the style of a co[m]pact theology. What this means is a problem of a general kind that I examined in another context. (See "Begriffsgeschichte als Philosoph[ie]," *Archiv für Begriffsgeschichte* 14 [1970]; *Gesammelte Schriften*, 2:77–91.)

20. Even Proclos recognizes the logical function of the "divine art," the[...] on the basis of an artificial interpretation of 31a (κατὰ τὸ παράδειγ[μα] δεδημιουργημένος). See E. Diehl, ed., *Procli Diadochi in Platonis T[imaeum] Commentaria* (Lipsiae, 1903), 133d.

21. Not only Krämer (see n. 19 above) but many others argue thus.

22. The extent to which the concept of life colors the concept of the v[...] can be seen further from Aristotle's summary passage in *Metaphys[ics]* 1024a11 f., where the ὅλον, the maimed, succeeds the κολοβόν, th[...]

23. Theophrastus, *Metaphysics* 5b.

CHAPTER 4
The Ontological Problem of Value

1. Hegel, *Phenomenologie der Geist*, ed. Hoffmeister (Leipzig and H[...] 1905), pp. 423 ff.: "Der seiner selbst gewisse Geist."

2. H. Lotze, *Kleine Schriften* (Leipzig, 1885–91), 3:305.

3. H. Lotze, *Mikrokosmos* (Leipzig, 1856–64), 2:416.

4. Ibid., p. 276.

5. Friedrich Nietzsche, *Gesammelte Werke* (Munich, 1920–29), 8:3[...]

6. Alois Roth, *Edmund Husserls ethische Untersuchungen: Dargeste[llt anhand] seiner Vorlesungsmanuskripte* (The Hague, 1960).

7. Ibid., p. 105.

8. N. Hartman, *Ethik* (Berlin, 1916), pp. 227 ff.

9. M. Scheler, foreword to 3d ed. of *Materiale Wertethik*; now in *Gesammelten Werken* (Munich, 1966), 2:21.

CHAPTER 7

The Ethics of Value and Practical Philosophy

1. See chap. 4 above.

2. See Alois Roth, *Edmund Husserls ethische Untersuchungen: Dargestellt anhand seiner Vorlesungs manuskripte* (The Hague, 1960).

3. Review of H. Schilling, "Das Ethos des Mesotes," rpt. in *Gesammelte Werke*, 5:300–3.

4. Theognis 435 ff.

5. J. Ritter, *Metaphysik und Politik: Studien zu Aristoteles und Hegel* (Frankfurt, 1969).

6. See my essay "On the Possibility of a Philosophical Ethics," chap. 2 above.

7. Gerhard Krüger, *Philosophie und Moral in der Kantischen Kritik* (Tübingen, 1931).

8. M. Scheler, *Der Formalismus in der Ethik und die materiale Wertethik*, in *Gesammelte Werke* (Bern, 1966), 2:20.

9. It is not sufficiently observed that not only the *elenxis* of the early Socratic dialogues, but even the dihairetic dialectic of the later ones, depend on anamnesis. This first became apparent to me in interpreting the *Sophist*. (See my lecture to the Philosophical Club in Washington, D.C., Fall 1981, in *Gesammelte Werke*, 7:338–69, and also 3:246 ff. and 404 ff.

10. I could also study the work that Ernst Tugendhat presented in Heidelberg on 2 Oct. 1980 under the title "Antike und moderne Ethik." It is now to be found in the proceedings of the Heidelberg Academy of Sciences, 1981, report 1. (The "ethics of material value" is not treated in this piece.) However close Tugendhat's work is to my own intentions in the essay "On the Possibility of a Philosophical Ethics" (see chap. 2 above), which he criticized, and however much in agreement I am with him (more than he acknowledges) in defending Kantian formalism, I would like to see more discussion in his own work about what "grounding" means in the realm of practical philosophy, and about the only thing it can mean if philosophers do not let themselves be pushed into the false role of experts or moral authorities. "A sharpened claim to grounding," which does not take into consideration the issues I develop above, seems to me unacceptable. (See also *Gesammelte Werke*, 3:333–49 and 350–74.)

11. N. Hartmann, *Ethik* (Berlin and Leipzig, 1926), p. 286.

12. K. O. Apel, *Transformationen der Philosophie* (Frankfurt, 1973), vol. 2.

13. Krüger, *Philosophie und Moral in der Kantischen Kritik*.

14. My inaugural lecture in Marburg has since appeared in revised form as "Friendship and Self-Knowledge: Reflections on the Role of Friendship in Greek Ethics" (chap. 9 in this volume).

CHAPTER 8
Reflections on the Relation of Religion and Science

1. Compare the essays "Mythos und Vernunft" and "Mythos and Logos," in *Gesammelte Werke*, 8:163–73.
2. See esp. the essay "On the Phenomenology of Ritual and Language," in *Gesammelte Werke*, 8:400–40.

CHAPTER 9
Friendship and Self-Knowledge: Reflections on the Role
of Friendship in Greek Ethics

1. B. Snell, *Aischylos und das Handeln im Drama*, Philologus Suppl. XX 1 (Leipzig, 1928).
2. See also my critique of E. Wolff's *Platos "Apologie"* (Berlin, 1929), in *Gesammelte Werke*, 5:316–26.
3. H. Langerbeck, ΔΟΞΙΣ ΕΠΙΡΥΣΜΙΗ. *Studien zu Demokrits Ethik und Erkenntnislehre, Neue Philol. Untersuchungen* H. 10 (Berlin, 1935). See my review in *Gesammelte Werke*, 5:341–3.
4. Cf. Dirlmeier's commentary on Aristotle's ethics in E. Grumach, ed., *Aristotles: Werke*, vols. 6–8 (Berlin, 1956–68). O. Gigon, "Die Selbstliebe in der Nikomachischen Ethik des Aristoteles," in A. Skiadas, ed., *Dorema* (Festschrift for Hans Diller) (Athens, 1975), pp. 77–114. J. C. Fraisse, *Philia: la notion d'amitié dans la philosophie antique* (Paris, 1974).
5. See also my essay "Vorgestalten der Reflexion," in *Gesammelte Werke*, 6:116–28.
6. *Nicomachean Ethics I* 4, 1166a1 ff.; *Eudemian Ethics H* 6, 1240a8 ff.; *Magna Moralia B* 15, 1212b33 ff.
7. *Nicomachean Ethics I* 9, 1169b3 ff.; *Eudemian Ethics H* 12, 1244b1 ff.; *Magna Moralia B* 15, 1212b24 ff.
8. See, e.g., *De anima A* 1, 403a6; *A* 3, 407b17; *B* 2, 414a20; *Γ* 9, 432a22 ff.
9. *Nicomachean Ethics I* 4, 1166a1; *I* 8, 1168b4; as well as *Eudemian Ethics H* 6, 1240b12.
10. I have never understood why people in general contrast the role of friendship in Aristotle to the role of justice in Plato. To me it seems clear that friendship with oneself is an artificial concept of Socratic origin when the expression "self-love" is used in a positive sense.
11. *Magna Moralia B* 15, 1212b34 ff., and *Eudemian Ethics H* 12, 1245b14.
12. See also *Politics H* 3, 1325b12 ff.
13. *Aleibiades I* 133a; *Phaedrus* 255d.

1. See my essays on the history and limits of the concept of value in this volume, chaps. 4 and 7.

2. *Nicomachean Ethics* Z 9, 1141b33, 1142a30; *Eudemian Ethics* Θ 1, 1246b36.

3. Many of my later works in volumes 5 and 6 of the *Gesammelte Werke* show this clearly. Since then I have lectured on my recent observations concerning Aristotle's conceptual formulations as well as here (on the occasion of a lecture I held in Gallarate, Italy). These works are follow-up on the foundational essay, "On the Possibility of a Philosophical Ethics," chap. 2 above.

4. G. Krüger, *Philosophie und Moral in der Kantischen Kritik* (Tübingen, 1931).

Essays were selected from the *Gesammelte Werke* (Tuebingen: J. C. B. Mohr [Paul Siebeck], 1986-) as follows. (Dates in parentheses are original dates of publication for the essays.)

INDEX

Allegory, 91

Anselm, Saint, 5, 14

Aristotle: 14; and the divine, 38, 41, 54-56; and ethics, 18, 28-30, 33, 106-8, 114-15, 142-61; and friendship, 132-41. *See also* phronesis; Plotinus

Atheism, 1, 119

Barth, Karl, 128

Beauty, 63-69, 84-85, 148

Being, 53, 55

Brentano, Franz, 69

Christianity: 59, 79, 91, 94, 119; and love, 141

Conscience, 25

Cornford, F., 39

Cults, 124

Democritus, 77

Dilthey, W., 66, 67

Divine, the: as being, 39; as life, 40, 47; as logos, 48; as thought, 47; as the whole, 38, 43

Duty, 23-24, 146

Eleatic school, 43, 47, 56

Emanation, 87-89. *See also* Plotinus

Enlightenment, 33, 82, 119-20, 158

Ethics: philosophical 18-36; Christian, 33

Ethos, 34, 75, 113, 115, 143, 155

Euripides, 16, 140

Faith, 121

Ficinus, Marsilius, 79

Friendship, 117-18, 128-41

God: 140; proofs of, 3-17

Good, the, 84-85

Gnosticism, 79, 89

Hartmann, Nicolai, 20, 26-27, 66, 71-72, 103-15

Hedonism, 148

Hegel, G.: 10, 20, 62, 94

Heidegger, M., 20, 127, 129, 153

Heyne, H., 92

Hölderlin, F., 17, 91

Holy, the, 127

Homer, 38, 81, 130

Humanism, 91

Husserl, E., 70, 151

Idealism, 12, 160

Jaspers, K., 16

Kant, I.: categorical imperative, 23, 61, 110, 142, 150, 158-59;

Yale Studies in Hermeneutics
Joel Weinsheimer, Editor

Yale Studies in Hermeneutics provides a venue for inquiry into
the theory of interpretation in all its varieties and domains.
Titles in the series seek to expand and deepen our
understanding of understanding while explicitly framing and
situating themselves within the tradition of recognized
hermeneutical thinkers from antiquity to the present.

• • •

Other volumes available from Yale University Press

Gadamer's Hermeneutics
A Reading of Truth and Method
JOEL C. WEINSHEIMER

Philosophical Hermeneutics and Literary Theory
JOEL C. WEINSHEIMER

Hermeneutics Ancient and Modern
GERALD L. BRUNS

Eighteenth-Century Hermeneutics
Philosophy of Interpretation in England from Locke to Burke
JOEL C. WEINSHEIMER

Introduction to Philosophical Hermeneutics
JEAN GRONDIN
Translated by Joel C. Weinsheimer,
with a Foreword by Hans-Georg Gadamer

Hermeneutics and the Rhetorical Tradition
Chapters in the Ancient Legacy and Its Humanist Reception
KATHY EDEN

Giordano Bruno and the Kabbalah
Prophets, Magicians, and Rabbis
KAREN SILVIA DE LEÓN-JONES

Rhetoric and Hermeneutics in Our Time
A Reader
EDITED BY WALTER JOST AND MICHAEL J. HYDE

Praise of Theory
Speeches and Essays
HANS-GEORG GADAMER
Translated by Chris Dawson

CPSIA information can be obtained at www.ICGtesting.com
Printed in the USA
LVOW08s0933191013

357607LV00001B/23/P